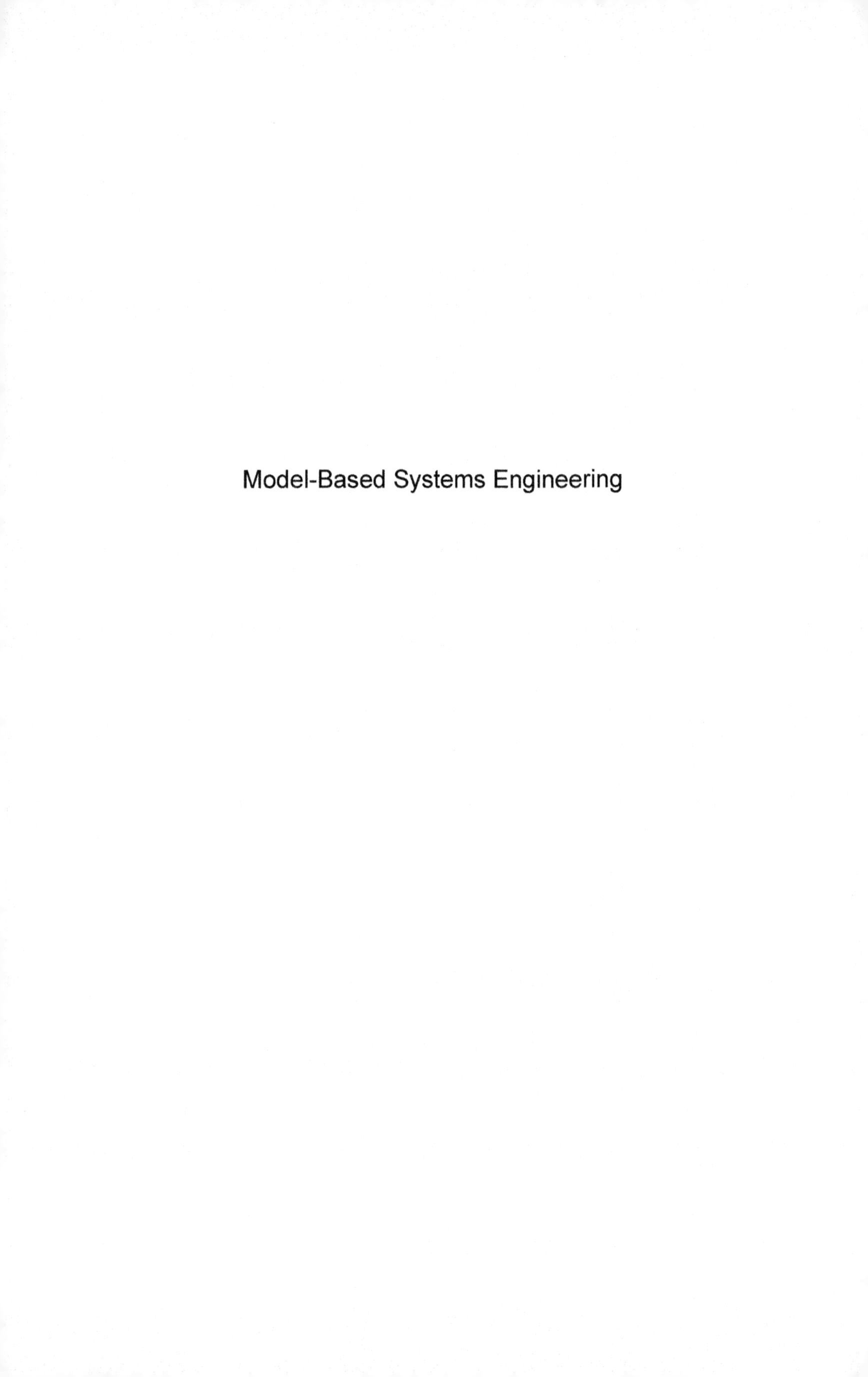

Model-Based Systems Engineering

Series Editor
Jean-Charles Pomerol

Model-Based Systems Engineering

Fundamentals and Methods

Patrice Micouin

WILEY

First published 2014 in Great Britain and the United States by ISTE Ltd and John Wiley & Sons, Inc.

Apart from any fair dealing for the purposes of research or private study, or criticism or review, as permitted under the Copyright, Designs and Patents Act 1988, this publication may only be reproduced, stored or transmitted, in any form or by any means, with the prior permission in writing of the publishers, or in the case of reprographic reproduction in accordance with the terms and licenses issued by the CLA. Enquiries concerning reproduction outside these terms should be sent to the publishers at the undermentioned address:

ISTE Ltd
27-37 St George's Road
London SW19 4EU
UK

www.iste.co.uk

John Wiley & Sons, Inc.
111 River Street
Hoboken, NJ 07030
USA

www.wiley.com

© ISTE Ltd 2014
The rights of Patrice Micouin to be identified as the author of this work have been asserted by him in accordance with the Copyright, Designs and Patents Act 1988.

Library of Congress Control Number: 2014945532

British Library Cataloguing-in-Publication Data
A CIP record for this book is available from the British Library
ISBN 978-1-84821-469-9

Contents

LIST OF FIGURES AND TABLE xi

ACKNOWLEDGEMENTS xvii

FOREWORD xxi
Dominique LUZEAUX

INTRODUCTION. GOALS OF PROPERTY MODEL METHODOLOGY xxv

PART 1. FUNDAMENTALS 1

Chapter 1. General Systems Theory 3

1.1. Introduction 3
1.2. What is a system? 4
1.3. Systems, subsystems and levels 9
1.4. Concrete and abstract objects 11
1.5. Properties 12
1.5.1. Material and formal properties 12
1.5.2. Accidental and essential properties, laws and types 13
1.5.3. Dispositions, structural and behavioral properties 17
1.5.4. Resulting and emerging properties 18
1.6. States, event, process, behavior and fact 20
1.7. Systems of interest 23

CHAPTER 2. TECHNOLOGICAL SYSTEMS 25

2.1. Introduction. 25
2.2. Definition of technological systems 25
2.2.1. Artificial autotelic and heterotelic systems 27
2.2.2. Technical-empirical and technological systems 27
2.2.3. Purpose of a technological system. 28
2.3. Function, behavior and structure of a technological system 30
2.4. Intended and concomitant effects of a technological system 34
2.5. Modes, mode switching and states 36
2.5.1. Modes of operation. 36
2.5.2. Mode switching 36
2.5.3. Operating states 37
2.6. Errors, faults and failures 37
2.7. "The human factor". 39

CHAPTER 3. KNOWLEDGE SYSTEMS 41

3.1. Introduction. 41
3.2. Knowledge and its bearers 42
3.3. Intersubjective knowledge 44
3.4. Concepts, propositions and conceptual knowledge. 45
3.5. Objective and true knowledge 47
3.6. Scientific and technological knowledge 50
3.6.1. Fundamental sciences 51
3.6.2. Applied sciences and technology. 53
3.6.3. Operative technological rules. 53
3.6.4. Substantive technological rules 55
3.7. Knowledge and belief 56

CHAPTER 4. SEMIOTIC SYSTEMS AND MODELS 59

4.1. Introduction. 59
4.2. Signs and systems of signs 60
4.3. Nomological propositions and law statements 65
4.4. Models, object models, theoretical models and simulation. 66

4.5. Representativeness of models and the expressiveness of languages 71
4.5.1. Representativeness of models 71
4.5.2. Expressiveness of a language. 73

PART 2. METHODS 77

CHAPTER 5. ENGINEERING PROCESSES 79

5.1. Introduction. 79
5.2. Systems engineering process 81
5.2.1. General framework 81
5.2.2. Design process 83
5.2.3. Safety assessment process. 88
5.2.4. Requirement and assumption validation 90
5.2.5. Verification of the implementation regarding requirements 91
5.2.6. Managing configurations. 92
5.2.7. Process (quality) assurance, certification and coordination with authorities. 93

CHAPTER 6. DETERMINING REQUIREMENTS AND SPECIFICATION MODELS. 95

6.1. Introduction. 95
6.2. Specifications and requirements 98
6.3. Text-based requirements and subjectivity. 100
6.4. Objectifying requirements and assumptions through property-based requirements 102
6.4.1. Definition. 102
6.4.2. Examples 104
6.4.3. Typology and sources of PBR 106
6.5. Conjunction and comparison of property-based requirements 110
6.5.1. Comparison of two PBRs. 111
6.5.2. Conjunction of two PBRs. 112
6.6. Interpreting text-based requirements. 114
6.6.1. Example 1: FAR29.1303(b) flight and navigation instruments. 115
6.6.2. Example 2: FAR29.951(a) Fuel systems – General 119

6.7. Conclusion: specification models and concurrent assertions 121

CHAPTER 7. DESIGNING SOLUTIONS AND DESIGN MODELS 127

7.1. Introduction. 127
7.2. Deriving requirements. 128
7.3. Basic system model of a type of systems 131
7.4. Dynamic design models of a type of systems 133
7.4.1. Behavioral design model (BDM) 133
7.4.2. Equation-based design models (EDMs) 139
7.5. Derivation and allocation of the system's behavioral requirements 141
7.6. Static design models 142
7.6.1. Composite system model 142
7.6.2. Structural design model 145
7.6.3. Allocation of BDM components to SDM components 146
7.7. Derivation and allocation of system requirements 146
7.8. The end of the design process and the realization 148

CHAPTER 8. VALIDATING REQUIREMENTS AND ASSUMPTIONS 151

8.1. Introduction. 151
8.2. The validation process according to the ARP4754A 152
8.2.1. Goal of the validation 152
8.2.2. Means of validation 154
8.3. The validation process according to the property model methodology. 156
8.3.1. Goal of the validation 157
8.3.2. Means of validation 158
8.3.3. Exactness of a system specification model 160
8.3.4. Validating the derivation of system requirements 161
8.3.5. Scenarios and validation cases, efforts and rigor in validation. 162
8.4. Conclusion. 167

CHAPTER 9. VERIFYING THE IMPLEMENTATION STEP BY STEP 169

9.1. Introduction. 169
9.2. The verification process according to the ARP4754A . . . 170
9.2.1. Goal of the verification 170
9.2.2. Verification methods 170
9.3. The verification process according to the property model methodology 173
9.3.1. Objects to be verified 173
9.3.2. Goal of the verification 174
9.3.3. Verifying the design 175
9.3.4. Verifying the other products of implementation 179
9.3.5. The contract theorem 181
9.4. Conclusion. 181

CHAPTER 10. SAFETY ENGINEERING 183

10.1. Introduction. 183
10.2. The safety assessment process according to the ARP4754A 184
10.2.1. Goal of safety assessment process 184
10.2.2. Means to assess safety 185
10.3. The safety assessment process according to the property model methodology (PMM) 191
10.3.1. Errors, faults and failures 191
10.3.2. FHA and interpretation of the 1309(b)(2)(i) requirements as PBRs 193
10.3.3. PASA/PSSA and deriving safety requirements 200
10.3.4. Simulation and validation of the derived safety requirements 204
10.3.5. Simulation and verification of the failure prevention mechanisms. 206
10.3.6. Reliability design models. 207
10.3.7. Safety theorem: validating additional requirements 208
10.4. Conclusion. 211

CHAPTER 11. PROPERTY MODEL METHODOLOGY DEVELOPMENT PROCESS 213

11.1. Introduction 213
11.2. Early phase of a system development, preliminary studies 213
11.3. Steps of the industrial development of a type of systems 215
11.4. Initial step: highest level system specification 216
11.4.1. Initial step general approach 217
11.4.2. Establishing a specification model of the type of systems 218
11.5. Design steps: descending and iterative design of the building blocks down to the lowest level blocks 226
11.5.1. Design step of a non-terminal block 227
11.5.2. Behavioral design step of a terminal block 229
11.5.3. End of the design step 231
11.6. Realization step of the lowest level building blocks 231
11.7. Integration and installation steps 232
11.8. Conclusion 233

APPENDIX 235

BIBLIOGRAPHY 253

INDEX 261

List of Figures and Tables

List of Figures

Introduction

I.1. ARP4754A engineering processes xxvi

Chapter 1

1.1. System composition, structure and environment 5
1.2. Concrete and abstract systems composition 6
1.3. Concrete and abstract systems environment 8
1.4. Representation of fictions (Arezzo Chimera and pi number) 12
1.5. Concrete and abstract objects 13
1.6. Laws and law statements. 21
1.7. Systems involved in the systems engineering processes. 23

Chapter 2

2.1. Artificial systems 26
2.2. Structure of technological system design 29
2.3. Function–behavior–structure framework 31
2.4. Function–behavior–structure relationships. 34
2.5. Failure, fault and error 38

Chapter 3

3.1. Conceptual knowledge classification 50

Chapter 4

4.1. Sign, concept and represented object 60
4.2. Rosetta stone, signification and meaning 62
4.3. Page extracted from Voynish manuscript 63
4.4. Sentence, proposition and fact 63
4.5. Signs and symbols 64
4.6. Law statement, nomological proposition and
factual law 65
4.7. Systems and models 67
4.8. Object model and theoretical model 70
4.9. Fault tree representation. 73
4.10. Reliability block diagram 74

Chapter 5

5.1. ARP4754A engineering processes 82
5.2. EIA632 system breakdown structure 83
5.3. EIA632 building block requirement
definition subprocess. 84
5.4. EIA632 building block solution
definition subprocess. 85
5.5. EIA632 building block design processes 87
5.6. ARP4754A safety assessment and
system development process integration. 88
5.7. EIA 632 design process extended to
safety assessment topics 89

Chapter 6

6.1. EIA632 building block requirement
definition subprocess. 96
6.2. EIA632 system specification architecture 97
6.3. Airfoil structural properties and
related PBR. 109
6.4. Interpretation: a process from
TBR to PBRs 114
6.5. Some PBRs linking inputs and outputs
of an air data computer 118
6.6. Some PBRs linking inputs and outputs
of an aircraft fuel systemE. 120

6.7. Canonical graphical representation of a system 121
6.8. System specification model. 124

Chapter 7

7.1. EIA632 building block solution definition subprocess. 128
7.2. Design model typology. 132
7.3. Basic system model. 133
7.4. Fuel system behavioral design model 134
7.5. System mode model 136
7.6. "Feeding the engines" functional chain 138
7.7. Basic system model including an equation design model. 140
7.8. Composite system model. 143
7.9. Composite system model. 144
7.10. Fuel system structural design model 146
7.11. EIA632 system design process. 148

Chapter 8

8.1. ARP4754A validation and verification processes connection 151
8.2. ARP4754A validation process model 155
8.3. Specification model tree validation. 157
8.4. System specification model exactification process by simulation 160
8.5. Subsystem specification model validation process by simulation 161
8.6. Graphical representation of a specification model 162

Chapter 9

9.1. Implementation verification process regarding specifications 170
9.2. ARP4754A verification process model. 171
9.3. Composite system model. 175
9.4. System model verification bench 177
9.5. System model integration verification bench. 178
9.6. EIA632-based system verification process. 180

Chapter 10

10.1. ARP4754A safety assessment process 185
10.2. Failure, fault and error 191
10.3. EIA632 requirement definition
process extended to safety aspects. 194
10.4. Interpretation of FAR29.1309(b)(2) and
systems FHA 195
10.5. Baro-altimeter specification model. 197
10.6. Fault-tolerant baro-altimeter specification
model. 200
10.7. EIA632 solution definition process extended to
safety aspects. 201
10.8. SDM virtual component computing
failure rates. 205
10.9. SDM virtual component
computing DAL 205
10.10. Reliability design model b) derived from
a baro-altimeter structural design model a). 207
10.11. Extended EIA 632 design process
model. 212

Chapter 11

11.1. PMM system development process 215
11.2. Colossus with feet of clay in
Nebuchadnezzar's dream 216
11.3. System specification model. 219
11.4. CAS specification model of intended functions.. 220
11.5. CAS protected volume and local environment
representation. 221
11.6. CAS specification model PBRs. 224

Appendix

A1.1. PMM workbench conceptual design 238
A1.2. Front-end main window and views on
PMM models.. 239
A1.3. PMM specification view 240
A1.4. Function and PBR graphical editor 243
A1.5. Equation design model view 245
A1.6. Behavioral design model view. 246
A1.7. Process graphical editor. 248

A1.8. Structural design model view 249
A1.9. System model and specification and design tree. . . . 251
A1.10. Component-block binding in system
model editor. 252

List of Tables

Chapter 1

1.1. Endo and exo-structure for concrete and
abstract systems 9

Chapter 6

6.1. Document-centric versus model-centric
paradigms. 102
6.2. Tolerances on computed altitude by an ADC
required by the SAE-AS8002A 116

Chapter 8

8.1. Validation case for a level of rigor
consistent with no safety effect (NSE) 164
8.2. Validation case for a hardened
level of rigor 166
8.3. Validation scenarios 166

Chapter 10

10.1. Failure rate classification according to AC29.1309.b(1) 185
10.2. Failure condition severity
definitions according to AC29.1309.b(2). 186
10.3. Safety objectives for
installed systems according to AC29.1309.b(3)(ii). 187
10.4. Top-level function FDAL
assignment according to ARP4754A. 188
10.5. Examples of system design
patterns considered for safety aspects 202

Acknowledgements

I would like to thank all those who contributed to this book, often unknowingly, during a reading, a discussion, a comment about a technological fact or rule and what not.

First and foremost, I would like to acknowledge the intellectual debt that I took from Mario Augusto Bunge, the Argentine-Canadian philosopher of science. His work, discovered by chance in [DUR 02], became to me over the years the most demanding, the most encyclopedic, the most fertile and, especially, the most true of contemporary epistemologies (because, of course, all epistemologies are not of equal value [BUN 12]). His vision, ontological, epistemological and methodological, is not only of a philosophical background, but also the tool that allowed me to develop an approach for the development of technological systems which is, both theoretically sound and practically effective and efficient. I have maybe sometimes misunderstood concepts and propositions supported by Mario A. Bunge, or I have reused them roughly, but his work and how he puts it forward are both a solid foundation on which to build and also an invitation to think for myself and, I hope, for the best. I also accept full repsonsibility for weaknesses. I would also emphasize the conceptual borrowings I took from Vladimir Hubka, Norbert Roozenburg and John Gero.

In second place, comes the Doctor, and Major General, Dominique Luzeaux. I had the privilege of working with him in the French chapter of INCOSE (AFIS). Chairman of this chapter, he showed me, practically, what could and should be a rational decision making process. He gave me the opportunity to take an interest in the systems of systems. I thank him warmly for having forworded my work, that I think innovative and thus exposed to controversies.

Then follow those who, at Airbus Helicopters (AH) and the Centre National des Etudes Spatiales (CNES), provided me with direct support in my research effort. They are Michel Palandri, Charles Lanzalavi, Louis Fabre, Pascal Pandolfi and Sebastien Poussard (from AH) and Erwann Poupard (from CNES). They allowed me not only to validate the concepts and approach initially imagined, but they also pushed me to deepen it. I owe them, first, for allowing me to formalize the notion of EDM and, second, for building a mock-up of property model method (PMM) front-end.

I would also like to thank all those whom I have rubbed shoulders with daily in recent years, during the development of the EC175, until its certification in January 2014, and now, on a new helicopter development. I think of airworthiness teams with O. Jeunehomme, D. Strutzer, G. Brun, A. Illinca and C. Bousquet, but also of the EASA counterparts such as P.G. Colombo, G. Soudain, A. Smerlas, C. Rosay and R. White, those of AH Design Office, M. Achache, A. Ducollet, A. Jenni, K. de Bono, M. Godard, S. Bailly, J. Nobili and M. Lanteaume, and those of laboratory test, G. Cahon and C. Gaurel and flight test, P. Bremond, A. Di Bianca, M. Oswald and A. Delavet.

I also think of those with whom I have exchanged ideas on these issues, more sporadically by the force of circumstance, but always with benefits: R. Miginiac and J.M. Lecuna from

Dassault Aviation, P. Farail and Y. Bernard from Airbus Avionics and Simulation Products, G. Meuriot and J.P. Daniel from AREVA, F. Hasse from DCNS, J.R. Ruault from DGA, E. Combes and J. Personnaz from PSA Peugeot Citroën, J.M. Pelbois from CMR, D. Seguela from CNES, and F. Malburet and P. Veron from Arts and Metiers Paris' Tech.

I would also like to thank my family, my daughter Louise, my son Pierre and especially my wife Monique who has been, at the same time, the most diligent, the most critical but also the most benevolent of my readers. I have needed them to be able to bring this project to fruition.

Foreword

With the ubiquitous spread of potentially disruptive technology in our daily life, we face continuously, increasingly complex systems, and as engineers it is legitimate to ask ourselves whether we are still able to manage that complexity, whether requirements and specifications, as well as their verification and validation are still needed and even possible. Patrice Micouin's book clearly answers "yes" and discusses a framework, both epistemological and practical, in order to address that challenge.

Patrice Micouin advocates a change of paradigm and introduces the Property-Model Methodology which relies on three claims:

1) in order to improve requirement engineering and aim for less ambiguous requirements that are better understood by the various stakeholders and the suppliers, a property-based requirement theory is needed;

2) in order to have objective specifications that can be interpreted by all stakeholders in the same way, model-based systems engineering should be used: the models are executable within simulations and yield results that are easily accessible and shared by the stakeholders;

3) simulation becomes the main means to validate the specifications and verify the proposed designs, but does not fully replace verification of the unitary physical products as well as their integration.

All existing systems engineering processes call for an initial effort on having from the start testable, measurable and unambiguous requirements. However, the methodology developed by Patrice Micouin goes further, as it aims at cancelling any interpretation margin in the understanding of a requirement.

Such a paradigmatic change relies on the capacity of expressing specifications as membership constraints on the properties satisfied by the system and recursively its components (when condition C is fulfilled, property P should take its values in the given domain D of numerical values). Restricting the class of specifications in such a way is justified by an epistemological discussion inspired by Mario Bunge's philosophy. Furthermore it should be noted that this class covers many real-world industrial applications and objectives in a much more immediate way than the behaviors obtained through simulation. Since such specifications build a semi-lattice, there is also a thorough way to go from system specification to component specifications, and back. On the one hand, this facilitates the transition from systems engineering to the different engineering disciplines needed to address individual components, and conversely. On the other hand, it avoids the usually tedious tasks of writing interface requirement specifications and interface definition documents.

As simulation is a common tool in many engineering disciplines, it is then much more straightforward to also use simulation as an effective tool at systems level, and engage in a paradigm transition from paper-centric engineering to model-based systems engineering. Practical efficiency considerations meet thus paradigmatic developments in this work, which is a landmark in the evolution of systems engineering practices that were formulated half a century ago when technology did not offer all the possibilities offered now.

Dr. Dominique LUZEAUX
Former Chairman of the French Chapter of INCOSE
July 2014

Introduction

Goals of Property-Model Methodology

I.1. Introduction

This book is an introduction to a systems engineering methodology, called property-model methodology (PMM). This compound noun is formed from two of its main characteristics, namely, the formulation of requirements due to the concept of property (property-based requirements (PBR)) and the adoption of a model-based systems engineering (MBSE) approach.

In this introduction, we will begin by (1) giving a brief description of this methodology, and will then present (2) its goals and (3) the processes undertaken and how these goals can be satisfied. In the last section, we will present the organization of the book.

I.2. Brief overview

PMM is a methodology for developing the following technological system types: discrete systems (such as avionics), continuous systems (such as fuel systems), mixed systems (such as electrical generating systems) and multiphysical systems (such as landing gear systems).

This is a descending development approach, arranged from top to bottom, but it authorizes the reuse of pre-existing blocks at all hierarchy levels of a type of systems.

This development approach is compatible with current industrial development standards, specifically ARP4754A and EIA632.

For example, it covers rigorously the following ARP4754A processes:

1) Development Process:

   a) requirement determination;

   b) architecture and design;

2) Requirement and Assumption Validation Process;

3) Implementation Verification Process;

4) Safety Assessment Process;

5) Configuration Management Process;

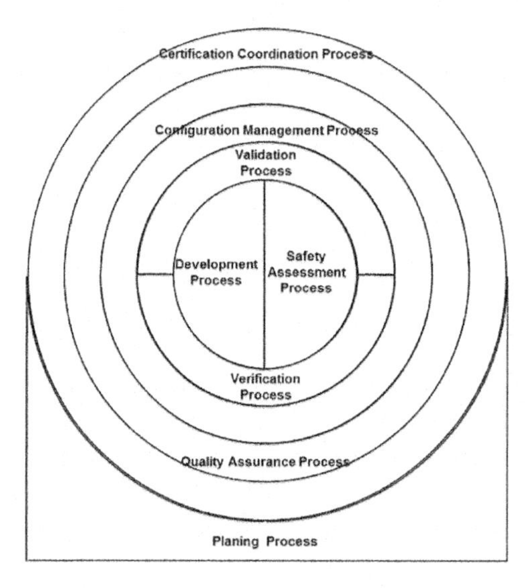

Figure I.1. *ARP4754A engineering processes*

whereas it supports the remaining processes:

6) Planning Process;

7) Process Assurance Process;

8) Certification Process.

Moreover, similarities can easily be drawn from other current standards such as those of the spatial or automobile sector.

This methodology [MIC 13] is based on three main pillars:

1) The first pillar concerns requirement engineering and the implementation of the theory of PBR [MIC 08].

2) The second pillar concerns the principles of MBSE, which more or less radically replaces the specification documents and design description documents by specification models and design description models[1]. However, this is an original MBSE approach, unrelated to the theoretical work of Wymore [WYM 93], and unconnected, either, with the huge literature on SysML [OMG 12].

3) The third pillar, simulation, is the primary means of validating specifications and verifying designs, while the verification of physical products, their integration and installation are maintained.

This methodology also includes a strict development process that, on the one hand, forces good practices and, on the other hand, eliminates poor ones. Therefore, it is quite the contrary of a toolkit supplied without a manual and that leaves users faced with a task, without guidelines, as most

1 "In particular, MBSE is expected to replace the document-centric approach that has been practiced by systems engineers in the past and to influence the future practice of systems engineering by being fully integrated into the definition of systems engineering processes", extracted from Systems Engineering Vision 2020, Document INCOSE-TP-2004-004-02, Version 2.03, p. 15, September 2007.

languages are presented (which is normal) but also as some MBSE methodologies (which is clearly abnormal).

Finally, it is based on modeling and simulation languages whose relevance and expressivity have been well demonstrated. The language VHDL-AMS [MIC 13] is currently the target language, but other languages, such as Modelica® [MOD 12], could be also target languages.

I.3. Goals

As all methodologies should claim it, the goal of PMM is to reduce the complexity of developing technological systems as well as to provide technological systems satisfying the needs, i.e. the right systems, in a timely manner and all for an objectively justified cost.

However, from current observations, many large projects concerning the development of technological systems do not satisfy these objectives: poor performances, problems with reliability after set up, significant delays or costs. Although none have been cited, there are many examples, for instance, in the aeronautical domain.

Among the causes, some of the most frequently mentioned are incorrectly introduced innovations, underestimated development times and not properly mastered outsourcing of design activities, which have opposite results to those intended (to reduce delays and costs).

We consider that all these causes have one common mode, namely, the poor quality of specifications communicated from the purchaser to the supplier (internal and external and at any level), while, on the one hand, the supplier still appears to trust overambitious subsystems and, on the other hand, the requirement specifications are the cornerstone of development process models in industrial contexts[2]. Therefore, just like a building that is transformed into a colossus with feet of clay, it begins to sway: misunderstandings and dysfunctions in development systems (systems made up of human beings) even more important than project organizations involve teams that are increasingly numerous and increasingly spread over very different cultural and linguistic areas.

Uncertainties about delays, costs and conformity of the produced system are the three observable symptoms of development process complexity[3], which PMM aims to reduce.

I.4. Processes

To achieve its goal, decreasing the complexity of developing technological systems, PMM introduces a methodological rupture as important as that of introducing the SCADE technology [BOU 14] in the development of software onboard aircraft.

This methodological rupture includes:

– the objectivation and the exactification of the specifications assigned to the systems to be developed;

– the design of solutions that are objectively error free;

– the delivery of specifications for subsystems objectively validated with respect to the system specification (i.e. consistent with the system level);

– the anticipation of approval phases of physical subsystems and of verification of integrated and installed systems;

2 Due to their starting options, Agile methods do not form part of this.

3 As defined by N.P. Suh [SUH 05].

– the reuse of equipment specifications and design descriptions when they are archived in libraries.

I.4.1. *Objectifying and exactifying the specifications*

Usually, the specifications are presented as text-based requirements (TBRs). Despite all efforts, they are still too interpretable and extremely dependent on the subjectivity of the author as well as the reader.

For example, it is quite difficult, from a pilot's perspective, to perform the validation of a long specification of several hundred boring pages, whereas using a test bench, he can immediately detect satisfactory, acceptable or prohibited situations.

To improve this situation, specifications must be objectified, i.e. rendered non-interpretable and understood in the same way by all stakeholders concerned. This is why PMM specification models can be performed on a simulation bench and the results of this simulation can be presented to the stakeholders concerned.

Being objectified, the specifications then become exactifiable; in other words, the stakeholders are put into a situation where they are able to decide on the validity of behaviors observed, just as they could do on a test bench. The deviations can be corrected until the specification is declared exact (as exact as possible).

I.4.2. *Designing error-free solutions*

When the description of the design of a type of systems is frozen in order to allow further development, it must be error free, i.e. it conforms to its specification. All possible execution pathways, all possible failures should be identified and all protective mechanisms and reconfigurations should be assessed. In other words, when a design is frozen, we should be able to rely on it. The level of confidence should be up to the hazards associated with the operation of the corresponding systems.

This is why PMM design models are simulated and why, when the requirements are breached, the simulator signals the effects of the errors to be corrected as well as where they are situated in the model. This simulation/correction process continues until no more requirements are breached. Afterward, the design model will be deemed error free, provided that the simulation effort was sufficient.

I.4.3. *Providing error free specifications of sub-systems*

When the specifications of subsystems of a system are provided to suppliers for the development of these subsystems, they should be entirely objective and as exact as the system specification itself.

This is the reason why, in the PMM approach, simulation allows us to establish whether the specifications of subsystems of a system are complete and consistent with one another, but also, with the specification of the system from which they derive, with the level of rigor required for the safety challenges of the system.

I.4.4. *Anticipating approval phases of physical units and their integration*

From the specification phases of individual components, the scenarios for testing the individual physical components should be obtained according to the verification effort required. This is directly obtained, in PMM, due to the particular form that assumes the expression of PBRs. The test cases are derived directly from the PBR structure. This allows an early estimate of the test means to be implemented as well as the verification costs of the individual physical components.

Similarly, when PMM is implemented, since the specification phases of systems and subsystems, the test scenarios and cases for system and subsystem integration are known according to the verification effort required. This allows an estimate of the integration and installation test methods to be implemented as well as the costs of the verification of the integrated and installed physical subsystems and system.

I.5. Conclusion

The remainder of the book is organized into two main parts and an appendix.

Part 1 introduces the underlying concepts of the PMM approach. It consists of four chapters (Chapters 1–4). Chapter 1 provides the framework for general systems theory (GST) and the subsequent three chapters cover the types of systems manipulated by designers of technological systems. The first chapter is about the technological systems themselves. The last two chapters (Chapters 3 and 4) are concerned with the means used by the designers to achieve their goals. On the one hand, this involves the knowledge systems used during an engineering process: factual knowledge and engineering reasoning and, on the other hand, the system of signs that allows the sharing of this knowledge among all stakeholders, including intermediate representations (models) of the targeted type of technological systems. These chapters provide the theoretical basis to the methodological proposals discussed in Part 2. If the readers wish to eagerly tackle this as soon as possible, they may skip the first part and move directly onto the second part. They will then be able to return, if necessary, to this foundation part (Part 1), whenever they feel that the ideas based on common sense deserve a theoretical explanation.

Part 2 describes the engineering process associated with the PMM. This part consists of seven chapters (Chapters 5–11). Chapter 5 provides a general framework inspired from the standards ARP4754A [SAE 10] and EIA632 [ANS 03]. Then the following five chapters describe the main categories of activities associated with this process (1) to determine the requirements and specification models (Chapter 6), (2) to define a solution and the design models (Chapter 7), (3) to validate the requirements and assumptions (Chapter 8), (4) to verify the implementation (Chapter 9) and (5) to perform safety analyses (Chapter 10). The last chapter of this part (Chapter 11) is dedicated to describing the development process and the different stages of development.

Finally, the Appendix describes the elements of a metamodel assuring that a system model is translated into a VHDL-AMS [IEE 08, IEE 07, ASH 03] simulation model.

PART 1

Fundamentals

1

General Systems Theory

1.1. Introduction

What do a nerve cell, the mathematical field of complex numbers and the Rosetta stone have in common? Nothing much, apparently. However, all three are systems, each in its own way. To grasp the unity behind this diversity of appearances, we must resort to general systems theory (GST), a theory that does not concern a specific type of systems in particular, but instead what makes a system a system. In the following, we will refer to GST, developed by Mario A. Bunge, in particular in volume 4 of his *Treatise on Basic Philosophy* [BUN 79]. We consider that Bunge's theory develops that of L. von Bertalanffy [BER 69], as well as renews it.

In this chapter, we define a system as a composite object characterized by (1) its composition, (2) its environment and (3) its structure. We will differentiate two types of systems: abstract or concrete depending on whether the objects composing the system are abstract or concrete. We will examine the relationships between the components and the environment according to whether the systems are abstract or concrete. We will also introduce the concepts of a subsystem and a level. More detailed analysis of the objects and properties will be done depending on whether the objects are material or abstract. We will introduce different classifications of properties: accidental and essential properties with the related concept of a type, structural and behavioral properties with the related concept of a dispositional property and, finally for systems, the resulting and emerging properties. We will also define the concepts of state, event, process, behavior and fact. The chapter will conclude with the three types of systems of interest for systems engineering: technological systems, systems of knowledge and systems of signs (or semiotic systems).

1.2. What is a system?

Following on from Bunge, a system Σ is an object composed of several parts (its components). These components have relationships between each other. We call endo-structure S_{int} of a system the network of these relationships between components, whereas the system components may have relationships with objects that do not form part of the system and what we call the environment E. The network of relationships between system components and the environment is called the exo-structure S_{ext} of this system. The structure S of a system is, therefore, the union of both its endo-structure and exo-structure: $S = S_{int} \cup S_{ext}$.

In summary, a system Σ is an object denoted by a triplet (C, E and S) such that:

– card(C) > 1, which expresses the fact that Σ is composed of several parts (composite object);

– $S = S_{int} \cup S_{ext}$ with $S_{int} \neq \varnothing$, which expresses the fact that the endo-structure of Σ is not empty.

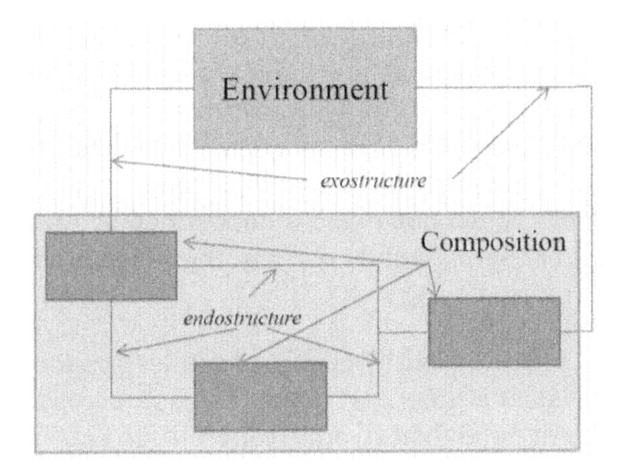

Figure 1.1. *System composition, structure and environment*

When E is empty, we say that the system Σ is a closed system. In the opposite case, we would say that this is an open system.

When its endo-structure S_{int} is empty, the object Σ is not a system; this is the case for fictitious stellar objects such as constellations. Ursa Major, unlike the galaxy M31 or the galaxy of Andromeda, is not a stellar system, whereas the stars that form it are themselves systems.

The definition of a system that we state deviates from the one proposed by von Bertalanffy and by many authors later on, beginning with the definition given in [SEB 13], namely "a system is a set of elements in interaction". In fact, in the case of the definition by Bunge, a system is an object whereas in the second case, it is a set. This may appear to be a negligible difference, two slightly different ways of designing the same reality. However, we claim at the opposite that the definition by Bunge provides us with a particularly fruitful characterization of a system, whereas its definition as a set prevents this characterization. In fact, a set is a particular type of object, i.e. an abstraction, resulting from a movement of mind, which allows different individual elements to be taken as a whole, whatever they are, (Ursa Major, for example). If all systems (concrete or abstract) are considered as sets (i.e. are mathematical beings), they are fictions resulting from movements of mind (brain processes), then systems could not exist independent of human beings who think of them. This is exactly the point of view held by constructivists[1].

Bunge provides an opposing realist vision to these constructivist theories: the objects exist according to two very different modalities: only concrete objects really exist objectively, whereas abstract objects only exist as fictions (which we will discuss later on in Chapter 3). So, a system may be either a concrete object (i.e. a material object) or an abstract object (i.e. a fiction) with all components of a material system being material objects, whereas in an abstract system it is only composed of abstract parts.

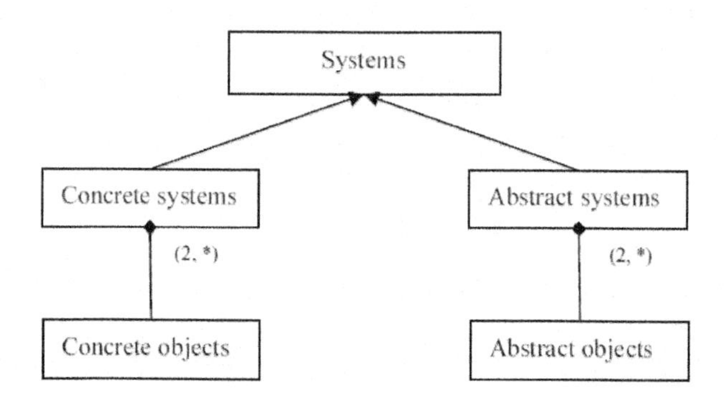

Figure 1.2. *Concrete and abstract systems composition*

1 Thus, Jean-Louis Le Moigne in his "General System Theory, Modeling Theory" (Dunod, 1984) recruits the physiologist Claude Bernard to support the constructivist theories (into the context of Claude Bernard's sentence: "systems are not in the nature, but in the mind of men", the word "systems" designates, unambiguously, intellectual constructs, theoretical and philosophical systems, while the existence of concrete systems is neither considered nor settled).

The following are examples of concrete (material) systems: the solar system, a macromolecule, the central nervous system of *Homo sapiens*, a family unit. Similarly, a project team, a book, a hospital, a factory and an airplane are also concrete systems, as well as a country's airspace, an energy production and distribution system at a continental scale. We are able to talk about the latter as systems of systems [LUZ 10]. To conclude with Bunge on this point, we hold the view that "the world is a world of systems" and that any concrete object is a system, a part of a system or both.

The following are examples of abstract systems: the mathematical theory of complex numbers field, the analytical mechanics of J.L. Lagrange, and the system of gods and goddesses of Olympus.

If, by definition, a concrete system is composed of concrete objects, however, it may have abstract systems in its environment; in this case, the concrete system is said to be capable of designating objects of abstract systems using concrete objects. Just as an example, languages, such as English and French (which are concrete systems), allow us to designate the same abstract concept of "system" in GST (which is an abstract system, more specifically a theory) using the different concrete words: "system" in English, "système" in French, "Система" in Russian, and so on'.

Similarly, if an abstract system is inevitably composed of abstract objects it may, however, have concrete systems in its environment; in this case, the abstract system is said to be capable of representing objects of concrete systems using abstract objects. Therefore, a theory such as GST (which is an abstract system) allows us to represent any type of concrete or abstract system and to point out the essential characteristics.

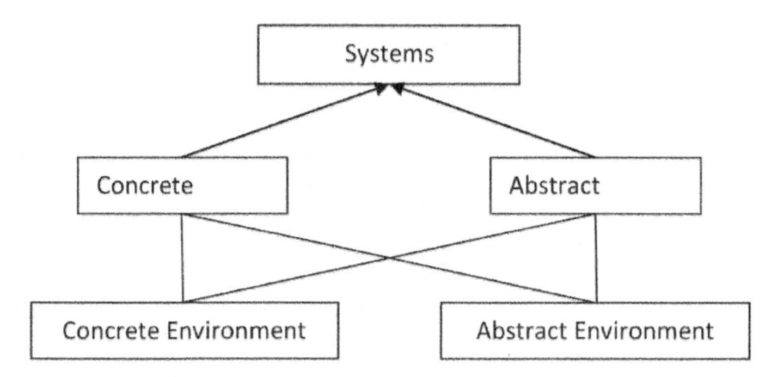

Figure 1.3. *Concrete and abstract systems environment*

The type of relationships, among the components of a system (endo-structure), on the one hand, and those connecting components with the elements in the environment (exo-structure), on the other hand, depends on the nature of the systems considered.

For the endo-structure, (1) with concrete systems, the relationships are material binding relationships (links) or non-binding material relationships; whereas with abstract systems, the relationships are formal relationships. We are concerned with links between concrete components when a change of state in some induces a change in state of others; this is typical in a mechanical system whose degrees of freedom are reduced by the links between parts. However, the links in question are not limited to mechanical links; they may be multiphysical (electrical, magnetic, nuclear etc.), chemical, biological or psychosocial. In addition to these links, the endo-structure may include non-binding material links such as topological, metric and temporal relationships. For example, "to be before, after, above, below, aligned with, centered, previous to, simultaneous, subsequent" are non-binding relationships. (2) For abstract systems, the relationships forming their endo-structure are formal relationships including logical operators (e.g. negation, conjunction and disjunction), relational operators (e.g.

equality, comparisons, belonging and inclusion), as well as assessment functions (e.g. "being well formed", "having the following meaning" and "being approximately true").

For the exo-structure, we find the same type of relationships as for the endo-structure; to these we must add relationships including those between concrete and abstract objects such as representation and designation relationships that will be discussed in Chapters 3 and 4.

	Concrete systems	Abstract systems
Endo-structure	Links Non-binding material relationships	Formal relationships
Exo-structure	*Concrete environment* Links Non-binding material relationships *Abstract environment* Designation	*Abstract environment* Formal relationships *Concrete environment* Representation

Table 1.1. *Endo and exo-structure for concrete and abstract systems*

1.3. Systems, subsystems and levels

The fact that a system is a composite object makes it possible for the components of a system to be considered separately. We can then ask the question: can the parts of a system be considered as systems for themselves?

We have an immediate answer to this question if we consider, for example, an atom of helium. This is definitely a system, composed of two electrons and a nucleus whose cohesion is assured by electromagnetic bonds (photons). Electrons are not systems since they belong to the set of elementary particles. The helium nucleus is also a system composed of two protons and two neutrons whose confinement is assured by strong nuclear bonds (gluons).

However, neutrons and protons are not considered as systems.

A system can therefore be composed of objects that are themselves systems. We can say that any systems composing a system Σ are subsystems of Σ.

If $\sigma \stackrel{\wedge}{=} (c,e,s)$ is a subsystem of $\Sigma \stackrel{\wedge}{=} (C,E,S)$, we obtain $c \subset C$, $E \subset e$ and $s \subset S$, which means that components of σ are components of Σ, the environment of Σ is included in that of σ and the structure of σ is included in that of Σ.

This possibility for a system to be composed of subsystems can obviously be repeated and this allows us to understand a system as being a hierarchy of subsystems with successive levels with the required decomposition depth.

For example, Dillinger [DIL 90] described a language as a system of signs (concrete system) that is composed of sentences, which are subsystems composed of clauses. These clauses are, in turn, subsystems composed of phrases. Phrases are subsystems composed of words, which are composed of morphemes, which, to finish, are systems of phonemes.

In other words, Dillinger provides us with a language description (true or not, this is another topic) as a system hierarchically organized into six levels (true or not, this is another topic):

1) sentences;

2) clauses;

3) phrases;

4) words;

5) morphemes;

6) phonemes.

1.4. Concrete and abstract objects

Starting from Aristotle's statement, from his *Metaphysics*, being (*ousia*) is a compound of matter and form (hylomorphism)[2]. It follows two complimentary ontological clauses: the world is exclusively made of concrete objects, and a concrete object is made of matter with material properties.

For example, according to the standard model of elementary particles, an electron is an elementary particle, characterized particularly by the following properties: a mass of 9.109×10^{-31} kg, an electric charge of -1.602×10^{-19} C, a radius less than 10^{-22} m and a spin of 1/2. Moreover, a photon is a stable particle with a spin of 1, and its electric charge and mass are zero. Assumed by G. Stoney, its existence was then proved by J. Thomson.

For Bunge, the hallmark of material objects would be to have energy [BUN 10]. In other words, energy would be a universal property of matter, whereas this would be lacking in immaterial objects, which is quite understandable given the incongruity of a phrase such as "the internal energy of the number π". As a corollary, the hallmark of concrete objects would be their aptitude for change, that is to say, they are capable of moving in a space of states. Briefly told, an object is concrete or material if and only if it possesses an energy or iff it is capable of being changed. Therefore, material objects have a real mode of existence, and according to both Bunge and Heraclitus "to be is to become".

According to this ontological assumption, concrete systems have energy and are able to change. This means that the composition C (t), environment E (t) and structure S

2 "By the matter I mean, for instance, the bronze, by the shape the pattern of its form, and by the compound of these the statue, the concrete whole" Aristotle, *Metaphysics Book Z*, Chapter III, translated by W.D. Ross.

(t) a specific system may change during the lifecycle of the system. Denotation of a concrete system Å (t)) by the triplet (C (t), E (t), S (t)) allows us to highlight its evolution over time.

On the contrary, an abstract object lacks energy and is immutable. It only exists as a fiction produced and reproduced by those who know it. The modes of existence of immaterial and material objects are, therefore, distinct. The transcendental number π, like a unicorn or a chimera, is an "eternal" object whose mode of being is that of fiction imagined by those who invented it or have knowledge of it, whereas the mode of being of the sun, galaxies and elements composing it down to the elementary particles is that of material reality, independently of any informed or non-informed individual.

Figure 1.4. *Representation of fictions (Arezzo Chimera and pi number)*

1.5. Properties

1.5.1. *Material and formal properties*

Just like the modes of existence of immaterial and material objects differ, their properties are also distinct. The properties of concrete objects are material or factual properties such as "having a position", "having a speed", "having an energy", etc. Properties would be considered as absurd if we tried to associate them with a concept or a proposition.

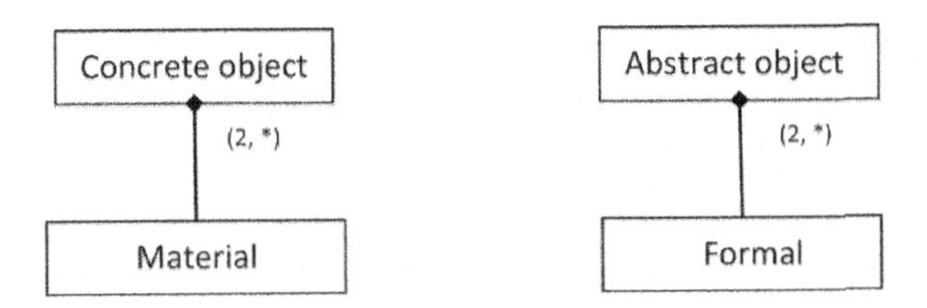

Figure 1.5. *Concrete and abstract objects*

Just like material objects, abstractions, such as a concept or a proposition, have properties called abstract or formal properties. Properties of abstract objects include the meaning of a concept or the truth of a proposition. Meaning and truth are properties associated with abstract objects, and it would be equally absurd if we want to relate them to concrete objects such as a stone or an airplane (signs, despite the fact that they are concrete objects, have a particular status).

1.5.2. *Accidental and essential properties, laws and types*

In this section, we are going to describe briefly the main assumptions and results from the theory of properties [BUN 77a, BUN 77b] designed by Bunge. Our theory of property-based requirements (PBRs) is based on these results, which will be discussed in Chapters 6, 7, 9 and 10.

When we consider objects and properties, there are two possible entries: an entry by the properties and an entry by the objects.

Using the first entry, the following definition is proposed: for a property P, the set of individual objects owning this property makes up the class C(P) of P. The class C(P) defines the extension of property P. For example, property E: "having an energy" for class C(E) represents the collection of material objects.

Bunge defines a precedence relation "≤" between the properties P and Q of material objects in the following manner: P ≤ Q if and only if C(Q) ⊆C(P); in other words, P precedes Q if and only if all objects owning the property Q also possess the property P. Thus, "to be a mammal" ≤ "to use a double articulation communication system" or even "to be an aircraft" ≤ "to be a helicopter" in the sense that humans who use double articulation communication systems are mammals, or even, helicopters are aircraft with rotating wings. According to the mathematical theory of sets, the relation "≤" defines a transitive and reflexive preorder [BOU 56]. However, this precedence relation is not antisymmetric since P ≤ Q and Q ≤ P do not imply that P and Q are identical, but only that P and Q are coextensive or concomitant, that is to say, the objects with P also possess Q and *vice versa*.

Bunge assumes that for two material properties P and Q, there is always a third material property R such that C(R)=C(P)∩C(Q), except if they are incompatible. He then defines the conjunction P∧Q of two material properties P and Q by posing C(P∧Q)=C(P)∩C(Q), as well as the incompatibility of P and Q by posing C(P)∩C(Q)=∅.

If we now consider the second entry mentioned above, the entry by the objects, we can define all the properties an object possesses and then can distinguish the essential properties of this object, on the one hand, and its accidental properties, on the other hand:

1) A property P of an object O is said to be essential if it is materially linked to other properties of O. For example, this is the case for the mass of Saturn's satellite called Pollux, which is linked not only to other static characteristics of its orbit (foci, apogee and perigee) but also to dynamic properties such as its period of revolution. Along with Bunge, we find that all material objects have essential

properties and each essential property of a material object has at least one link with one of its other essential properties.

We name a material law any relation L materially linking together essential properties $\{P_1, P_2, .., P_n\}$ of a material object O. When two or several essential properties $\{P_1, P_2, .., P_n\}$ of a material object O are linked by a material law L, the evolution of one of these properties will induce a change in one or several properties that are linked to the first, according to law L. We also note that the material law L is also a property of the material object O.

It is on this assumption that science and technology are based, depending on which essential properties of objects are linked together by material laws. If we remove this assumption, the project of science and technology becomes insane.

Taking essential properties of an object O into consideration will allow us, as we will see below, to define a space of states that is the same for all objects sharing the same essential properties.

It also allows us to define a type of objects. We define a type of objects as the collection of objects that share the same essential properties or even have the same real space of states. So, we can define the satellite type as the type that collects all the actual and potential satellites, the human type as the type collecting all the human beings (dead, alive or to come), the aircraft type, etc. like this.

Here, the term “type” is also quite general and can be specialized by introducing the concepts of “gender”, “species”, “order”, “kind”, etc., as systematics specialists do. For example, in the field of aeronautics, we can introduce the aircraft gender, and within type we can distinguish at least two species: fixed-wing aircraft (or airplane) and rotary-wing aircraft (or helicopter). Within the helicopter species, we can distinguish category A and category B kinds, according to helicopter regulations. Finally, within these kinds, we can distinguish types of helicopters. One type of helicopters is a collection of helicopters that have been or will be produced, according to the same type design definition (TDD), which has gained approval (type certificate (TC)) from the competent airworthiness authority, attesting that this definition conforms to the airworthiness regulations.

2) A property P of an object O is said to be accidental if it is not materially linked to any other property of O. As an example, this is the case for the color of aircraft flight recorders, familiarly called "black boxes", and which could be of any other color, if the regulation did not require, in a fairly conventional way, that flight recorders be orange.

If we then consider all individual objects with the name "Pollux", we define a fiction, a heterogeneous set of individuals including a hero from Greek mythology, a satellite of Saturn, a grammarian of the Greek language of the 2nd Century AD, a character of a dog from a television series, an elephant from Jardin des Plantes, etc. Note that naming a satellite of Saturn or an elephant in Jardin des Plantes "Pollux" does not change much, or even anything about the existence of this natural satellite or elephant. "To have the name Pollux" is not a material property of this satellite or elephant. The name of object materials, when they have one, is a property of the object represented within a system of signs used to denote it whereas this denotation has an accidental character, if we refer to the theses of linguist F. de Saussure on "the arbitrary nature of the sign" [SAU 00].

1.5.3. *Dispositions, structural and behavioral properties*

If we consider, for example, the electric charge of an electron, it is a typical characteristic of this object, an intrinsic property, whereas its velocity is a relational property of this object with regard to other objects forming a framework (for example, the frame of an instrument, a building and three fixed stars). However, in both cases, a property is inevitably a property of an object or a group of objects.

Flying is a property of aircraft, for example. This does not mean that an aircraft permanently flies but only that it is capable of flying, it has a disposition to fly. An aircraft spends most of its time on the ground in parking lots or on runways. During the periods spent on the ground, the ability to fly is retained, without being used. In other words, flying is a potential property or a disposition of aircraft that is only used at certain times, when they fly effectively.

As Roozenburg [ROO 91] states, manifesting a disposition requires conditions of actualization. For example, whether an aircraft has fuel or not, and whether there is a pilot to perform the takeoff or not, forms parts of the conditions of actualization in order for an aircraft to fly.

Conditions of use $\wedge$ *structural properties* $\rightarrow$ *actualization of dispositions*

However, object dispositions do not come from the conditions of actualization that reveal them, but rather from inner and sometimes hidden characteristics (structural properties) of the object. The object can do what it does, in given circumstances, because it is structurally how it is.

Therefore, an aircraft obtains its ability to fly from its wing (fixed or rotary). Flying and possessing wings are two essential properties of aircraft, linked by a material law, which we know from a lift law statement such as: $F_Z = 1/2\, \rho V^2 S C_z$.

The characteristics of a wing, such as its reference area S and its lift coefficient C_Z, are structural properties of the wing while the lift effect of a wing (exerted with force F_Z) is a behavioral property (or disposition) that is only actualized when there is a transfer to the surrounding atmosphere at a minimum velocity V.

1.5.4. *Resulting and emerging properties*

Classifying properties as resulting or, on the contrary, as emerging is typical with systems, unlike preceding classifications that were about concrete or abstract objects, without questioning whether they are systems or not.

For a property P of system Σ denoted by the triplet (C, E, S), P is said to be a resulting property of Σ when parts of Σ already possess the property P.

For example, for the property "have a signification": the words "system", "is", "object" and "composite" all have a signification, just like the statement "a system is a composite object" also has a signification. We can say that the signification of the statement above is a resulting property since its components already had one.

However, we can say that the statement "a system is a composite object" is true (or false), whereas saying that the word "system" is true sounds like an incongruity. This is due to the fact that a statement can have a value of truth, whereas a word cannot; "to be true", "to be approximately true" and "to be false" are possible emerging properties of a statement because the words that form it do not possess this property.

This is true for language, as well as for any type of systems: a passenger airplane has the emerging property of transporting people from one airport to another through the air, a property that no component of this airplane possesses.

Some authors associate the concept of emergence with that of complexity. For them, the emergence of properties within systems is related to their complexity. We do not share this point of view. For us, any system, whatever it may be, whether simple or complex, always has emerging properties. We can see this by simply considering the properties of a water molecule compared to those of its constituent atoms or even observing the incredible diversity of shapes (geometric, rose, cardioid, nephroid, limacon of Pascal, lemniscates, epicycloid and conchoid) that can be obtained using basic mechanisms: rolling circle without sliding over a base circle, or segment and circle. Thus, emergence is a distinctive property of the structure of systems without any particular association with complexity: for an object, the phrase "be a system" is equivalent with the phrase "possess emerging properties".

As Bunge points out, the emergence [BUN 03] of properties at the level of a system is also accompanied by the submergence of the components' properties, i.e. the properties of the components also disappear at the level of the system. Thus, the explosive nature of the sodium atom is submerged in the molecule of salt.

The emergence of properties can be considered in another sense, related to the previous: in the evolution of a line of objects, new properties may arise in the descending objects that were not there before. It is in this sense that we can say that human language emerges in communities of hominids who did not have language before. We could also question whether properties emerge within a line of technological objects. Also, it is in this sense that the modern helicopter emerged into gyroplanes at the beginning of the 20th Century. This question is not answered here.

1.6. States, event, process, behavior and fact

In the previous section, we mentioned that abstract objects are immutable, and assigning them a state is no more relevant than wanting to assign them a material property.

However, material objects (and consequently concrete systems) are changing by nature, that is to say that different material properties may change value. This includes the fact that composition, structure and environment of a concrete system change over time. More basically, we assume that material properties can be split into two subcategories: qualitative and quantitative properties. Qualitative properties are characterized by finite domains of values, whereas quantitative properties are characterized by infinite (countable or uncountable) domains of values.

If we take object O, characterized by its essential properties $\{P_i\}_{i \in I}$, and if we assume that each property P_i has a domain of values D_i (finite or infinite), then at each moment, each property P_i possesses a value p_i within domain D_i (for quantum properties, it is necessary to consider a distribution of values). Note that the value p_i of P_i may be constant (specific property) or variable (generic property).

We can then define the state e of an object O, at a given time t, as the set $\{p_i\}_{i \in I}$ of values of each of its essential properties, i.e. $e = \{p_i\}_{i \in I} \in \prod_{i=1}^{n} D_i$. Here, $\prod_{i=1}^{n} D_i$ that designates the space of theoretically possible states of O sets a reference framework. However, the material links between the essential properties of a material object determine the set of states $\{p_i\}_{i \in I}$ that are actually possible, whereas others are materially impossible. We call the actually possible space of states E of object O, or briefly, the space of states of O, the strict subset of $\prod_{i=1}^{n} D_i$ such that if $\{p_i\}_{i \in I} \in$ E then $\{p_i\}_{i \in I}$ is actually possible. Only, space E of the actually possible states is to be considered.

To illustrate the above problem, we can take the following example: a gaseous object G can be characterized by three generic properties: p pressure, v volume and T temperature, whose respective domains can be represented (in a system of units) by intervals $[p_{min}, p_{max}]$, $[v_{min}, v_{max}]$ and $[T_{min}, T_{max}]$ of the set of real numbers R. These three properties of object G are essential properties of G because they are linked to each other by a material law (whose effects can be observed using numerous experimental devices); in other words, the space of state E_G of G is a strict subset of the Cartesian product $[p_{min}, p_{max}] x [v_{min}, v_{max}] x [T_{min}, T_{max}]$.

Note that this material law that links these properties of G is itself a property of G. It is an independent property of the object, irrespective of whether we (a knowing subject) know it or not.

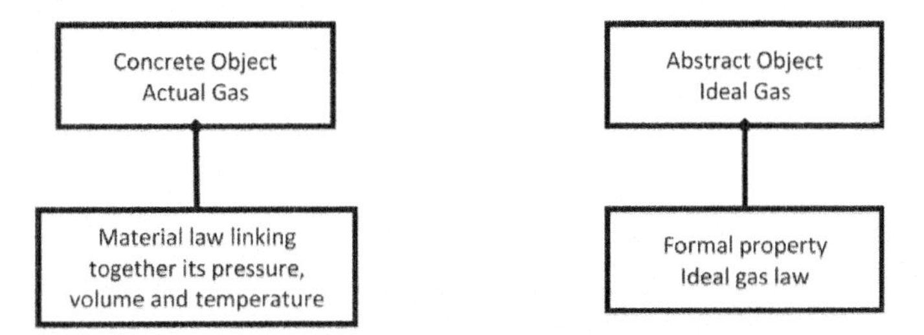

Figure 1.6. *Laws and law statements*

This material law must not be confused with law statements that are only approximately true representations of this material law: ideal gas law, van der Waals law, etc.

A material law, whether causal or stochastic, *actually* links the essential properties of an object with each other. This is what a law statement designates; for example, the ideal gas law designates the material law linking pressure, temperature and volume of the gas capsule.

The ideal gas law statement designates a nomological proposition, that is to say, an abstract object whose truth is only approximate in the sense that it imperfectly represents a material property of real gases, namely the interdependence of pressure, temperature and volume.

Given this, we can define an event as a transition from one state of an object to another (thus, a concrete object possesses at least two states, due to the fact that a concrete object is mutable). Moreover, a process is defined as a sequence or a continuum of events leading an object from an initial state to a final state.

Therefore, the emission or absorption of a photon is an event during which an electron "jumps" from one energy level to another, whereas the propagation of a nerve impulse is a process through a network of neurons. We must remember that an event, or a process, is inevitably an event or a process within an object, and that events or processes without a material support object are not real but abstractions.

Next, we can define *a* real behavior b (actually possible) of an object O as an actually possible trajectory in its actually possible space of states.

$b: t \in [t_0, t_1] \rightarrow s \in E$ where E is the actually possible space of states of object O.

In the same way, we can define *the* behavior[3] of object O as the set of actually possible behaviors of O.

To conclude, we introduce the concept of fact as follows: a fact is either a concrete object in a given state or an event (or a process) occurring in a concrete object.

For example, an aircraft flying in cruising level or a computer powered on are facts. The facts have to be distinguished from statements that denote them, "the aircraft is flying at cruise level" or "the computer is powered up".

1.7. Systems of interest

We say that systems engineering is concerned with three types of systems. Figure 1.7. shows the types of systems that will be involved in the systems engineering, either as its purpose (technological systems) or as the means to be used (knowledge systems and signs systems

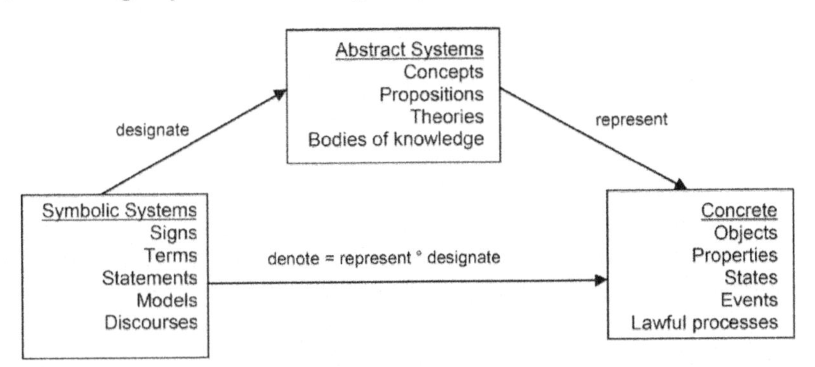

Figure 1.7. *Systems involved in the systems engineering processes*

3 Bunge calls it "mechanism". The behavior does not necessarily have observable effects. This is the case for static systems that have no observable reactions, to some extent, under the action of external forces.

First, and rather obviously, technological systems form the first type of systems considered in systems engineering. For example, the standard ISO15288, page 52 [ISO 02], indicates that "systems considered in this International Standard are man-made, created and utilized to provide services in defined environments for the benefit of users and other stakeholders". We dedicate one chapter (Chapter 2) to the characterization of technological systems as a type of concrete systems.

However, technological systems are not the only systems considered in systems engineering. A second category includes systems of knowledge and, in particular, scientific and technological systems of knowledge. In fact, as we will see later on, technological systems are artificial systems that are designed, produced, operated, maintained and dismantled by taking into account the available scientific (basic and applied) and available technological knowledge. We dedicate one chapter (Chapter 3) to the characterization of these systems of knowledge that are abstract systems.

Finally, a third category of systems is also considered in systems engineering, to which belong all documents, frameworks and all models used to design, produce, operate, maintain and dismantle technological systems, that is semiotic systems. It includes a lot of systems of signs which, on the one hand, denote the technological systems of interest and, on the other hand, designate the systems of knowledge representing the technological systems of interest. We dedicate one chapter (Chapter 4) to characterizing systems of signs and models (that form a subcategory of sign systems).

2

Technological Systems

2.1. Introduction

In this chapter, we will present technological systems as concrete objects with general characteristics of all concrete systems, as well as characteristics that are specific to them. Technological systems will be characterized as human creations, designed and developed on the basis of a usefulness (i.e. not for themselves but with an external purpose) and available scientific and technical knowledge. This will allow us to introduce, alongside general concepts of property, behavior and state, specific concepts of technological systems such as function, concomitant effect, mode, error, fault and failure that are necessary to define engineering processes of technological systems.

2.2. Definition of technological systems

The standard ISO15288 [ISO 05] defines the systems of interest as systems produced by humans, created and used to provide services in defined environments to benefit its users and other stakeholders. This definition states that systems with which systems engineering is concerned are artificial systems, designed and produced by humans. For this reason, these are systems orientated toward a determined objective (to provide services in defined environments to benefit its users and other involved stakeholders[1]). This, therefore, introduces two important differences compared with natural systems, whose genesis and dynamics are linked to the evolution of the universe and to which it is impossible to assign an aim ("no plastic brain, no teleology" [MAH 97]). As emphasized by Bertalanffy, there are two possible and equivalent scientific descriptions to the evolution of concrete systems: the first one by the initial and boundary conditions, and the second one by the end conditions [BER 69] (for example, due to the Maupertuis principle). This is true for all concrete systems, including biological systems [MAH 97]. The reason why one is preferred over the other is only due to convenience, even if the second has extra-scientific connotations.

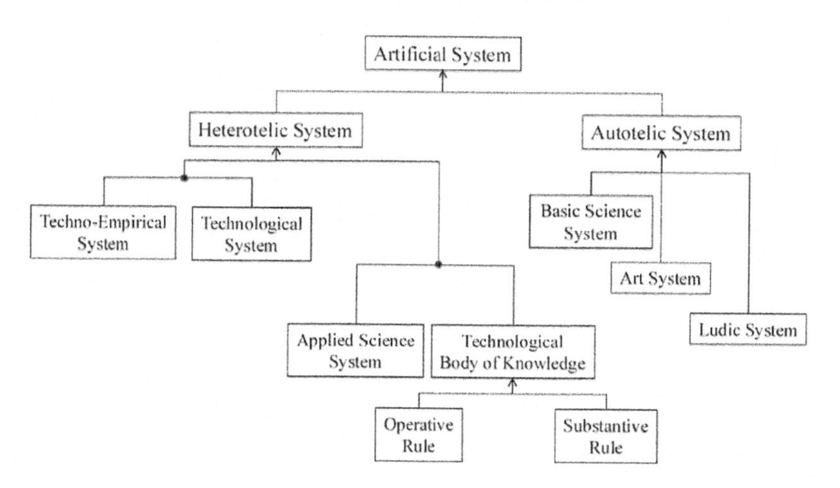

Figure 2.1. *Artificial systems*

1 An enterprise, organization, or individual having an interest or a stake in the outcome of the engineering of a system. Examples of stakeholders are acquirer, user, customer, manufacturer, installer, tester, maintainer, executive manager, project manager, and all other personnel having a stake in the development or outcome of the engineering of a system. The enterprise as a corporation or agency and the general public are also stakeholders.

2.2.1. *Artificial autotelic and heterotelic systems*

Among the artificial systems that humans can create, we can see that the systems of interest, mentioned in ISO15288, have a purpose, just like any artificial system; but in this case, this purpose is external to the system itself.

Autotelic systems: in other cases, this purpose is the system itself. For example, the purpose of a system of signs, such as Homer's *Odyssey* or the theory of functions of the complex variable of G. Riemann, is only the system itself. In particular, this is what has led us to consider that basic sciences, like some other cultural productions, were public goods [MER 42] (at the opposite of antiscientific theses such as "science is politics by other means"[2]). This includes autotelic systems, i.e. whose purpose (teleology) is the system itself, such as works of art, games or fundamental sciences.

Heterotelic systems: however, the systems of interest mentioned in ISO15288 are heterotelic systems, i.e. whose purpose is external to the systems themselves.

2.2.2. *Technical-empirical and technological systems*

However, one feature of system of interest seems to have escaped the ISO15288 definition, namely the systems with which systems engineering is concerned, and ISO15288 in particular, are not any heterotelic systems. ISO15288 rightly excludes technical-empirical systems (but does not justify this exclusion), that is to say, empirically based technical systems produced since ancient times by humans. As the philosopher Alain stated, "the technique (is) a kind of thinking that is applied to the action itself and is improved

2 LATOUR B., *Science in Action, How to Follow Scientists and Engineers through Society*, Harvard University Press, 1987.

by continuous trial and error"[3]. Hydraulic systems, such as Roman aqueducts with siphons, described by Vitruvius[4], have been constructed by engineers with incomparable expertise and experience, before Heron of Alexandria[5] discovered the principles of mass and momentum conservation that are included in the technological and non-empirical design of this type of work.

Unlike empirical systems, technological systems with which systems engineering is concerned, are systems that are, as Bunge emphasizes, systems designed and produced *conforming* to law statements and laws of applied science and contemporary technology. It is, therefore, possible to develop space systems, to evaluate their behavior in environment in which they will have to evolve, and to launch them into space with a reasonably low risk of failure. In other words, technological systems are materializations of applied sciences and technology, i.e. "a state of capacity to make, involving a true course of reasoning" [ARI 99].

2.2.3. *Purpose of a technological system*

If we now consider the purpose of a technological or an empirical-technical system, we find that this purpose is to change the state of the world or, in other words, to modify the environment of the desired system to be build. As N. Roozenburg noted, operating a technological system (or an empirico-technical system) consists of modifying the state of its environment according to the purpose of the system (which is external). For example, if this is an aqueduct, the

3 ALAIN, *Humanités 1946 in Les passions et la sagesse*, Gallimard, La Pléiade, pp. 292–294.

4 Vitruvius, Roman architect born around 90 BC and died around 20 BC, deals with the construction of aqueducts and syphons in *De architectura*, written around 25 BC.

5 Heron of Alexandria: Greek engineer, mechanic and mathematician of the 1st Century AD

purpose is to bring water to a town without water (by transporting it from a distant place where it is abundant). If it is a civilian aircraft, the purpose is to transport people or merchandises by airways from one location to another.

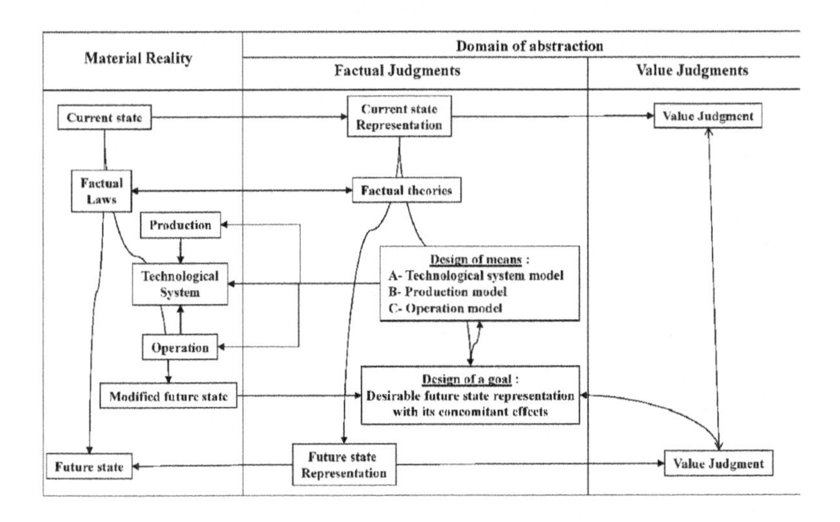

Figure 2.2. *Structure of technological system design*

Inspired by a figure presented by Roozenburg and Eeekels in their work [ROO 91, p. 11], and adapted above, we represent the design process of a type of technological systems as below.

Figure 2.2 is presented in three columns. The first column denotes facts (things in a given state, event and process) in the reality, whereas the other two columns are in the abstraction domain (in other words, in the head of different stakeholders). In reality, a current state objectively exists: for example, the presence of an obstacle (for an aircraft) that crosses the trajectory of another aircraft in evolution. This current state will evolve into a future state due to the laws of flight. In this example, the future state corresponds to a collision between these two aircraft. This scenario is not imaginary; it has occurred on several occasions, sometimes leading to catastrophic accidents. Retrospective analysis (abstraction domain) showed that both aircraft passed from a risk situation (judgment of value in a representation of the current state) to a catastrophic situation (judgment of value in a representation of the future state) inexorably given by the aircraft trajectories (physical theories). These accidents have caused the federal aviation administration (FAA) to require devices to be installed on board aircraft helping them to avoid these types of collisions (design of a goal: modified future state) whose solution (design of means) was studied for several years, i.e. a type of technological systems: collision avoidance systems (CAS, transponders in Mode C) that have been produced and installed on board transport aircraft (CAS) and light airplanes (transponders) and that the pilots must learn to operate.

As made explicit by Roozenburg, designing a type of technological systems involves two tasks:

– designing a goal;

– designing means to achieve that goal.

This will be presented in more detail in Part 2 of this book.

2.3. Function, behavior and structure of a technological system

In Chapter 1, we proposed the definition of concrete systems that does not refer to the concept of function, whereas the term "function" is systematically considered as a central concept of engineering and is regularly defined as an "intended effect of a system, of a sub-system of a product or a component" [SSE 14].

If this is so, it is because the systems perform functions for users who give them a use value. This also provides them with an exchange value and, therefore, can be included in an economic system.

In fact, we assert that the concept of function does not belong to the theory of concrete systems overall but only to that of heterotelic systems. Generally speaking, a concrete system $\sum$ does not have a function and natural systems, in particular, does not have functions. The Sun, for example, does have a structure (nucleus, radiation and convection zones in peripheral layers and photosphere) and a behavior (nuclear fusion, convection and radiation to the surroundings), but it does not have a function even if it has an effect upon all the components of the solar system and beyond. It is somewhat imprecise to say that the "function" of the nucleus is to produce energy that will then be radiated by the peripheral layers (photosphere). Natural systems can have interesting or valuable effects, but this is just a coincidence, a golden opportunity.

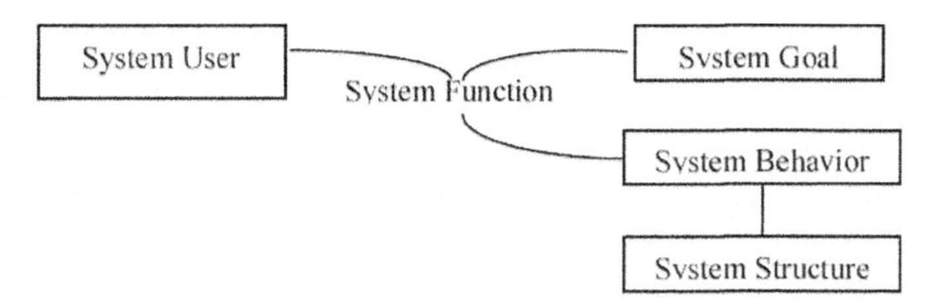

Figure 2.3. *Function–behavior–structure framework*

Conversely, heterotelic systems are designed and developed to produce an intended effect in its environment by exploiting a behavioral property of the designed system. This intended effect produced by the behavior of the system and embedded in its structure by its designer forms its function. This is what Hannah Arendt expressed rightfully, when she stated that hitting a nail with a shoe does not make the shoe a hammer [ARE 61], unlike Dori, who considered the function “hammer” of a stone (natural object) in his main book [DOR 02].

To define a function, we will use the conceptual framework proposed by John S. Gero, designed according to function–behavior–structure (FBS), and defined by Gero [GER 90] as:

1) *Function (F)* of an object is defined as its teleology, i.e. “what the object is for”.

2) *Behavior (B)* of an object is defined as the attributes that are derived or intended to be derived from its structure (S), i.e. “what the object does”.

3) *Structure (S)* of an object is defined as its components and their relationships, i.e. “what the object consists of”.

Therefore, we say that the functions of a system are the effects (intended by its stakeholders) of the interaction of this system with its environment (as Chandrasekaran and Josephson [CHA 00] defined for the function of a device).

One function of a system is the relation between a technological system, its designer who assigns it a purpose and its users who use the system to achieve the desired objective.

ARP4754A (p. 8) is a good example of this, with its definition of a function: “intended behavior of a product based on a defined set of requirements regardless of implementation” (while mentioning the possibility of detecting an unintended function – page 65 – is an antilogy).

The FBS design framework proposed by J. Gero, which distinguishes between the intended behavior B_e of a system Σ, which is the behavior of Σ, conceived to satisfy the intended functions, and the effective (and emergent) behavior B_s of structure S, which is the set of actually possible behaviors, taking into account the structure S of system Σ and the environmental conditions in which Σ is situated.

Figure 2.4 shows the existent relations between:

– the function F of a system Σ and the designed behavior B_e of Σ to satisfy the function F;

– the structure S of Σ designed to perform this intended behavior B_e of Σ;

– the structure S of Σ and the emergent behavior B_s of S of Σ;

– the comparison between the intended behavior B_e of Σ and the emergent behavior B_s of S;

– the function F of a system Σ and the emergent behavior B_s of S.

It is noticeable that no direct relations exist between the functions and the structure, which means that the structure S never directly carries out functions. This is what the "no-function-in-structure" principle expresses (originally by Kleer and Brown [BRO 84]) and is adopted by the research community named "Engineering Design", of which John Gero is a member.

The "no-function-in-structure" principle results from the following consideration: the structure of an object never directly performs an intended function of this object, whether an electronic video camera or the windscreen of an aircraft. In all cases, the device (structure), whether it is a windscreen or a camera, behaves in such a way that it provides an image of the source by processing it. It is the behavior of the object that performs the function, and this behavior is made possible by the object structure (under determined conditions of use). Therefore, it is the structure of an aircraft (airframe) that determines the way it behaves when subjected to flight loads and ground loads, and what makes it to perform its integrity function[6].

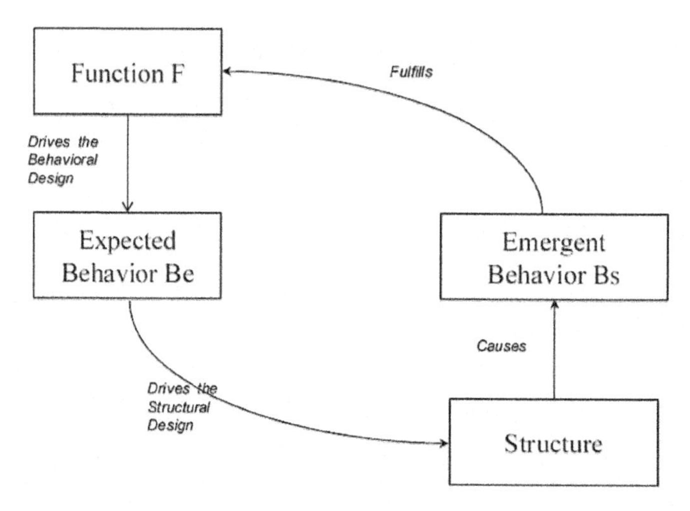

Figure 2.4. *Function–behavior–structure relationships*

2.4. Intended and concomitant effects of a technological system

However, we cannot reduce a technological system to its intended effects i.e. its functions (in fact, its usefulness). This is an important and surprisingly ignored fact[7] in many functional analysis manuals.

In fact, technological systems also produce non-desirable effects that we call concomitant effects or even intolerable effects. This is based on the fact that the essential properties

6 The integrity function of a system Σ is often performed by a static subsystem of Σ.

7 Note that Vladimir Hubka, who in his theory of technical systems (p. 48) introduced the concept of secondary output to account for undesirable effects, is an exception to this generality.

of a system are linked with each other by laws, as we have seen in Chapter 1. Thus, mobilizing one of these essential properties of a system, in order to produce an intended effect, inevitably leads to the mobilization of other properties linked to the previous one, which is not necessarily desired.

Using an electric motor, to drive a load, can concomitantly cause the motor and its environment to heat up, eventually reaching unacceptable temperature ranges.

Similarly, to produce heat from nuclear fission reactions, of course, will allow producing electricity downstream, which is the function of a nuclear power plant, but will also concomitantly produce some undesirable effects, which will be prevented and confined.

Thus, the engineering activity cannot be reduced to find an efficient solution to the limited problem of providing the intended effects of a system, but, equally, consists of finding an acceptable solution to the problem caused by the concomitant effects of this system.

To conclude, the intended and concomitant effects are not defined once and for all. If we consider, for example, a conductor carrying an electric current, we name the Joule effect, the heat radiation phenomenon induced by the electrical flow through the conductor. Basically, this effect is neither desirable nor undesirable, but its appeal depends on the context in which it occurs. In the case of a house equipped with electric heaters, the Joule effect is intended, desired by the occupants who want to warm themselves up; in the case where electrical energy is transported in electric cables, the Joule effect is quite on the contrary an undesirable effect, a loss for the operator.

2.5. Modes, mode switching and states

In this section, we define what is meant by a mode, a clearly distinguished concept of that of a state, as well as the relation that they maintain with each other. The aim of this is to clearly provide the basis for a logical representation of a system.

2.5.1. *Modes of operation*

The operation of a technological system during a time interval $\Delta\tau$ may be divided into exclusive modes of operation during which distinct purposes (functions) are assigned to the system.

In a system, each mode of operation has a specific corresponding behavior, which allows it to meet the objectives associated with this mode. For example, an aircraft automatic pilot, when engaged, functions in different modes during which it carries out specific objectives such as heading holding mode (HDG), altitude holding mode (ALT), etc. A purpose (function) may be specific to a mode; it may also be common to several modes, which then differ by other more specific objectives: its submodes.

2.5.2. *Mode switching*

A system switches from one mode to another by mode switching. The modes switching can be validly represented using Statechart diagrams, a graphical and typical hierarchical representation of states and transitions between states, provided by Harel [HAR 98]. However, note that, unlike transitions of supposedly instantaneous in the Statecharts representation, the switching between modes often involves complex operations, whose duration cannot be considered as negligible, and which should be constrained in terms of time.

2.5.3. *Operating states*

By the operating state of a system at a given instant, we mean the value of one or several properties of the system at this instant. For example, we can define the state of balance of an aircraft as being situated in its XY plane. The aircraft is said to be centered when the projection of its center of gravity in XY remains in a polygon established during its design.

For a system to function in a given mode, it may need to be in a particular state; so, mode of operation is linked to the state of operation. Inversely, in a given state, it may not be able to function in a given mode. In this second case, the mode and state of operation are, therefore, incompatible. For example, before an aircraft can take flight, the pilot has to check that it will remain centered (state) during all flight stages (takeoff, climbing, cruising, descent and landing modes) according to the payload (constant) and the fuel load (variable).

Describing a system in terms of states or modes is a dual description. Different operating states of a system correspond to different operating modes, which are mutually involved, or on the contrary, excluded.

For instance, for a fuel system to function in "Feeding" mode, its filling status will be higher than a certain level corresponding to the level of unusable fuel during flight. Similarly, for it to function in "Refueling" mode, its filling status will be lower at a certain level (in theory, full). In the opposite cases, the modes and states are incompatible.

2.6. Errors, faults and failures

When a technological system operates in environmental conditions for which it was designed, a failure occurs when the intended effect is no longer produced on its environment, whether it involves the loss of function or an erroneous functioning.

Moreover, if we analyze the space of states of a technological system, it can be divided into two regions. The first region includes the subspace of well-functioning states, i.e. those for which the technological system has all of its resources and capacities, behaves satisfactorily and consequently provides the intended services.

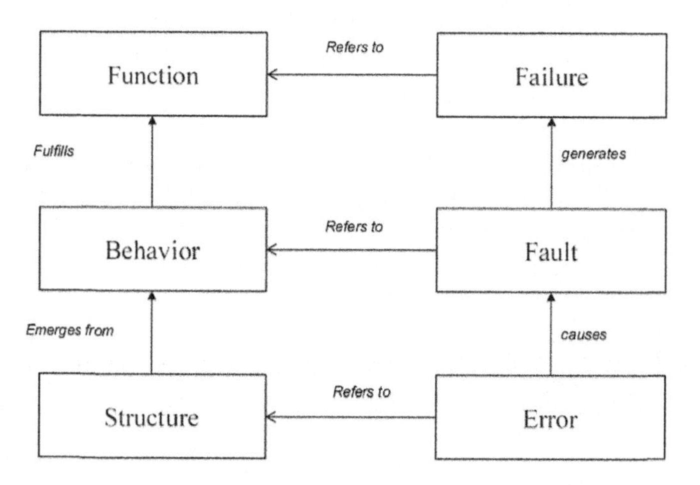

Figure 2.5. *Failure, fault and error*

Inversely, the second region includes the subspace of faulty states (states of dysfunction), that is to say, those for which the technological system does not have all resources or all capacities. It then behaves in a faulty manner, but this may not necessarily be visible, according to the external intended behavior. A faulty behavior is a behavior that temporarily or definitively includes faulty states. It is then unable to provide the services intended of it in a more or less short-term.

In agreement with the FBS framework adopted above:

1) A failure, situated at level (F), is the more or less immediate consequence of faults, i.e. the faulty behaviors of the system – situated at level (B).

2) Faulty behaviors, situated at level (B), originate from (human) errors, situated at level (S), whether in the structure of a technological system or in its composition, or when assessing the environment conditions in which it will operate.

We will describe these aspects in more detail in Part 2 of this book (Chapter 10, section 10.3.1).

2.7. "The human factor"

A technological system may at some point or another in its lifecycle include a human component. This is the case, for example, for aircrew (pilot and copilot) and the passengers of an aircraft when in flight. The crew not only assure the control and monitoring of an aircraft but they also participate, just like the passengers do, in its composition and properties, in particular, of its weight and centering.

These human components of a technological system are characterized by the following:

– intrinsic (anthropometric) and relational (biomedical) properties that provide them with biological capabilities, but also limitations;

– perceptive, cognitive motor, and rational but also affective properties that provide them with psychological capabilities, but also limitations;

– properties regarding exchange, communication, sharing and understanding, but also of conflict, that provide them with sociological capabilities, but also limitations due to belonging to multiple social groups, which implies that social interactions are governed partly by social values and norms;

– an often rational behavior (that is to say, a behavior that is adapted to situations that may arise and to the objectives pursued) but not always;

– a capacity to create (that is to say, a capacity to establish relationships between previously unlinked entities).

We consider that these multiple and complex characteristics should make human engineering the focus of technological systems, whereas these disciplines still remain insufficiently integrated.

3

Knowledge Systems

3.1. Introduction

In this chapter, systems of knowledge are introduced as a second pillar of systems engineering. In the previous chapter, we characterized technological systems as artificial systems with two essential traits: (1) they are developed and operated to provide services to persons (this is their purpose) and (2) they are designed and produced using the scientific and technological knowledge needed to ensure that this goal will be achieved. Thus, systems of scientific and technological knowledge form one of the key levers which allow engineers to carry out their task of developing technological systems that meet defined goals. Systems of knowledge are therefore the central part of technological systems engineering. This is why we will define, in the following, what we understand by scientific and technological systems of knowledge, and how this knowledge is integrated into the process of systems engineering.

3.2. Knowledge and its bearers

We assert that knowledge does not exist separately, nor does motion[1]. What exists are individuals who have knowledge, just as there are bodies in motion. This will be our initial assumption. In other words, in this chapter and throughout this book, knowledge does not refer to any world of ideas, such as the one attributed to Plato by a long philosophical tradition or even the one proposed by K. Popper who, in [POP 72], postulates the existence of a third world, i.e. that of objective knowledge, along with physicochemical phenomena (world 1) and that of knowledge, essentially subjective psychic activity (world 2).

We are not adopting the idea that knowledge might reside in books, models or any other form, such as databases of knowledge or other "ontologies" which are the objects of study for knowledge engineering experts [RUS 13]. In fact, we claim that these artifacts are representations which have value only for groups of individuals sufficiently educated to understand them; otherwise, they are completely worthless. Without Champollion and Egyptologists, the Rosetta stone would only be a beautifully crafted block of granodiorite bearing no historical knowledge. We will deal with the connection of knowledge with systems of signs in Chapter 4, which is dedicated to semiotic systems and in particular to models.

If individuals are the only bearers of knowledge, where does this knowledge reside? Along with neuroscientists, from D. Hebb [HEB 49], J.P. Changeux [CHA 97], G. Rizzolati [RIZ 05] or M. Gazzaniga [GAZ 09], we will say that the brain is the organ of knowledge. To know is one of the brain's dispositions, with that of feeling emotions (love, hate and

1 Knowledge as Motion are reifications, i.e. errors of treating as a concrete thing something which is not concrete, but merely an idea (A.N. Withehead).

fear). To be more specific, the neocortex is the part of the brain (80% of the human brain) whose plasticity allows us to "engram" knowledge.

This disposition to know is actualized through learning during which the individual often interacts with others, particularly during early learning (the child learns to walk, speak with relatives), even if, according to Spelke [SPE 07], we assume that infants have an innate capacity to acquire core knowledge. During this learning phase, networks of neurons will be formed, which neuropsychologists have called "engram" [LAS 50], "cell assembly" [HEB 49] or even "pexgo" [BIN 76]. These maps are the material basis of cognitive processes whose purpose is to acquire knowledge by shaping a previously uninformed or differently informed neuronal system: walking, playing violin, demonstrating or memorizing a theorem are, therefore, embodied cognitive processes. These cognitive processes are the activity of these systems of neurons during their formation, when knowledge is acquired, and then formed, when this acquired knowledge is used. The repeated summoning of knowledge reinforces bonds inside these systems of neurons. It also facilitates the adjustment or reorganization of this knowledge by the association and reorganization of neuronal systems due to the neocortex plasticity. Inversely, lack of use may lead to the unlinking of these neuronal systems, that is to say, forgetfulness. Damage to these neuronal systems may also lead to a loss of knowledge, such as in the case of aphasia [LUR 76].

This biological underpinning of knowledge allows us to propose the following classification:

– sensorimotor knowledge, such as walking, dancing or even playing the violin, knowledge that largely involves the sensory and motor regions of the neocortex;

– perceptive knowledge, such as observing and recognizing the song of a nightingale, the smell of jasmine or the configuration of a constellation in a starry sky, which engage the sensory regions (temporal for hearing, prefrontal for smell and occipital for sight) and the associated cortex (parietal lobe);

– conceptual knowledge, such a knowing a poem, knowing the quaternion theory or that of gas kinetics, which involve prefrontal regions of the neocortex.

Therefore, knowledge is a biopsychological object, due to their location and the biochemical nature of the processes involved. We can refer to individual or subjective knowledge, since they are first "engrammed" by the individuals involved. This knowledge evolves throughout the life of the individual. If $K_i(t_1)$ is the total knowledge of an individual, i is instant t_1; then at instant t_2, $K_i(t_2)= K_i(t_1)+AK_i(t_2-t_1)-FK_i(t_2-t_1)$, where AK_i and FK_i represent the knowledge acquired and forgotten during the interval t_2-t_1.

When we refer to a given piece of knowledge, we can abstract its result by equating the multiplicity of cognitive processes activated by an individual at different moments or several individuals corresponding to this piece of knowledge.

3.3. Intersubjective knowledge

Due to the way they are acquired, pieces of knowledge are also social objects. In fact, early learning (procedural and conceptual) for a very young child occurs through high emotional closeness to relatives; further learning, whether at school or work, occurs most often when in contact with teachers or peers. This interaction with teachers or peers allows the individual to adjust personal knowledge to that of others, to refine them, correct them and confirm them in a continuous validation process. Researchers have already described these mutual learning phases, during which stakeholders, of a company or a team, form a distributed cognitive system and adjust their subjective knowledge and produce shared or intersubjective knowledge [HUT 96, AUV 09]. Since G. Rizzolati discovered "mirror" neurons in different animals, and more recently in humans, we consider that this intersubjectivity also has a neuronal basis. These "mirror" systems play a key role in empathic phenomena which are a condition favoring the learning process.

In this neurosociological framework, we can refer to knowledge of a given social group S such as a family, a village, a business team [ENG 98], a company [SIM 91] and, more broadly, a society as the total knowledge shared by the individuals in this system.

$$K_S(t) = \bigcap_{i \in S} K_i(t)\,;\, Ks(t) \subset Ki(t)$$

3.4. Concepts, propositions and conceptual knowledge

Conceptual knowledge (know-what) is of particular interest to us within technological systems engineering, even though procedural knowledge (*savoir-faire* or know-how) has a key role in tuning phases in the development of a system.

The elementary piece of conceptual knowledge, referred to in the statement "S knows p", is the proposition designated by the statement p. A proposition, itself, is an abstract system composed of concepts[2], which are the components of any construct (propositions, theories, bodies of knowledge, etc.), just like elementary particles form the basis of any concrete object. Concepts may be extra-logical concepts: constant, such as that designated by the sign "Pollux" or collections such as that designated by "satellite" which form

2 It is at the expense of a word game, on which is a concept, that the C-K theory could arrange concepts and knowledge into two distinct spaces [HAT 03].

predicates such as those designated by "is a satellite" or "belong to the category of satellites", or logical concepts such as those designated by "not", "all", "entails". Thus, the proposition designated by the statement "Pollux is a satellite of Saturn" consists of two constant concepts designated by "Pollux" and "Saturn" and a predicate designated by "is a satellite of". Logical concepts form the endo-structure of a proposition. Its environment as a system provides a context to the proposition and its exo-structure is made up of all the connections that the proposition holds with the objects in its environment. Surrounding objects may be other constructs (concepts, propositions, theories, bodies of knowledge, etc.) but also the concrete objects and facts to which they refer. In particular, in scientific or technological propositions, connections to concrete objects and facts are designated as "semantic assumptions". When a proposition does not refer to any concrete object or fact, it is a formal statement and, in the opposite case, it is a factual statement (about facts).

The meaning M of a concept c or of a statement p, in a given context, has two components: first, its extension, i.e. the total referents R (all concrete and abstract objects that are referred to), and, second, its intension or content C, namely, the set I of propositions implied by c (or by p) and the set P of proposition implying c or p.

In a defined context, a proposition may or may not have a meaning. It is meaningful when its intension is not empty (i.e. with antecedents or consequents), whereas it is meaningless when its intension is empty (i.e. without antecedent and consequent). As an example, the proposition "the specification S is validated"[3] is meaningful in the context of the ARP4754A standard since it is connected with propositions such as "a requirement is validated when it is sufficiently correct and complete" or even "a specification S is

---

3 Refer to the second part of this book, Chapter 5.

a set of requirements and assumptions related to a system type Σ", whereas the following one "the system type Σ specified by S is validated" is meaningless in the same context.

Now, if we consider a context C including the proposition p, the context Citself forms a system whose composition is a set of propositions. These propositions are linked together by relationships such as inference relationships. This system has an environment composed of other systems such as formal systems: logical, mathematical, etc. The scientific and technological systems are characterized by the richness of their interdependencies (open systems), whereas pseudo-science and the magical thoughts are distinguished by their self-sufficiency and isolation *vis-à-vis* other areas of knowledge (closed systems).

3.5. Objective and true knowledge

Among the conceptual knowledge shared by a social group, some cannot be subjected to any conceptual verification process (e.g. a formal demonstration) or empirical verification process (e.g. an experiment). This is a type of knowledge that cannot be corroborated or falsified according to Popper [POP 02].

Quite the reverse, a piece of objective knowledge is a piece of knowledge that is able to be corroborable or falsifiable through a verification process that can be conceptual (a formal demonstration) or empirical (an experiment). The use of test apparatuses allows[4] or would allow[5] us to carry out verification procedures. The result of this verification will either confirm or refute the concerned piece of knowledge.

4 This is the case of the interferometer of Michelson.

5 This is the case of the VIRGO gravitational wave detector near Pisa, Italy.

When an insufficient amount of evidence is only collected, an objective piece of knowledge remains as an assumption[6].

Thus, propositions designated by statements such as "the Earth is round", "when the volume v is constant, the pressure p of a constant quantity of gas is proportional to its absolute temperature T" or even "light propagates in a medium called the luminiferous ether" are objective knowledge, in the sense that verification devices have been defined, which can either confirm or refute this knowledge when put to the test. This is, for example, the case for experiments carried out by Charles and Gay-Lussac which confirmed Charles law that linked the pressure and temperature of a gas at a constant volume; however, the Michelson and Moreley experiment refuted this assumption of the luminiferous ether. Therefore, intersubjective knowledge can be classified into two categories: objective or unverifiable knowledge. Systems engineering cannot rely on unverifiable knowledge.

Finally, some objective knowledge may be said to be true, such as the proposition designated by: "in the field of complex numbers, any algebraic equation of *n*th degree has exactly *n* solutions"[7] or even "the Earth is round"[8]. However,

---

6 This is the case for gravitational waves, whose existence was predicted in 1918 based on Einstein's general theory of relativity and, for which, impressive experiments are still ongoing (VIRGO and laser interferometer gravitational-wave observatory (LIGO)) or under preparation new gravitational wave observatory (NGO).

7 Gauss provided evidence of its truth in the form of four distinct demonstrations.

8 Its truth is based on observations and experiments, first the observation of moon eclipses, considered by Aristotle (*On the Heavens*) and us, as a valid proof of the rotundity of the earth. Other evidences followed with Eratosthenes, the Abbasid Caliph Al-Ma'mun of Baghdad, and in modern times with Magellan's circumnavigation or the Apollo 11 trip around the moon, after an obscurantist episode in which some religious scholars, such as Lactance or Cosmas of Alexandria, imposed literal readings of the Bible to fight a science in loss of audience.

these two statements are not equally true. The first statement, which is involved in an abstract system, is true in formal terms; it is a truth of reason (or a formal truth), whereas the second statement is related to a concrete object. The truth, associated with this type of propositions about real things and facts, is a truth of fact[9], which comes by degrees. The Earth is only round as a first approximation and the associated proposition is only partially or approximately true.

Establishing the formal truth of an abstract proposition or the factual truth of factual proposition is not done in the same way. The first is based on coherence, the coherence of the proposition, for which we wish to establish the formal truth with respect to its premises, whereas the second is based on the consistency of the factual proposition with the corresponding facts.

With regard to the factual truth of a proposition, it comes from the concordance between this factual proposition, its antecedents and consequences, with the corresponding facts. Thus, an individual has a true perception of a circular figure if, in fact, he/she perceives a circle when a circular figure is presented. An additional piece evidence was provided, when observing the activity of certain neuronal systems, an experimenter (neuroscientist) determined, rather precisely, what was perceiving the individual ([KRE 00], cited in [BUN 10]). Moreover, the consistency of cognitive processes with associated facts is only approximate. Consequently, we can assign only a partial truth value to a factual proposition.

9 The difference between truth of reason and truth of fact was introduced by Liebniz (Theodicy, 1710).

3.6. Scientific and technological knowledge

Bunge [BUN 10] inserts the systems of scientific and technological knowledge into the following classification:

– Knowledge may be illusory or, quite the reverse, genuine. When genuine, knowledge may be ordinary or specialized. Specialized knowledge can be divided into non-scientific or scientific.

– Systems of scientific and technological knowledge are systems of objective knowledge developed within research systems, i.e. social systems, formed by researchers sharing bodies of knowledge, problems, objectives and methods, including the scientific method considered as the general method for problem-solving. These research systems produce confirmations and refutations of current and new theories that enrich the initial bodies of knowledge [RAY 03]. These confirmations and refutations are made in accordance with the scientific method[10].

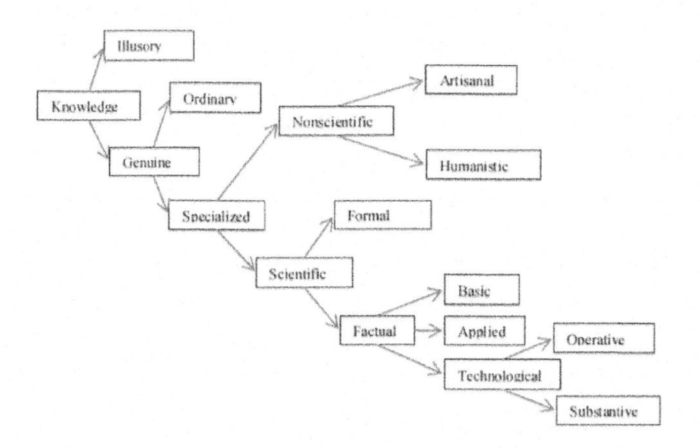

Figure 3.1. *Conceptual knowledge classification*

10 Thus, the demonstration of G. Perelman for the Poincaré's conjecture has been declared valid by the mathematical community after multiple verifications by his peers.

Depending on whether the purpose of this research is the only constitution of knowledge or another utility, these theories fall under fundamental sciences (autotelic systems) or applied sciences (heterotelic systems).

3.6.1. *Fundamental sciences*

The ultimate purpose of factual fundamental sciences is to discover, in the form of nomological propositions, the laws that link together the facts of a domain. The scientific approach involves defining the concepts that identify the essential properties of concrete systems (physical, chemical, biological, psychic and sociological) to represent, in the form of assumptions, laws that link these properties together. These assumptions are organized into theories, that is to say, in the form of hypothetico-deductive systems which, using basic postulates, derive theorems that can be deduced logically. These assumptions are then corroborated or refuted as scientific laws (nomological propositions) via a verification process which establishes whether a nomological proposition is or is not consistent with the referred facts. The verification process includes tools, protocols and observational and experimental data.

Depending on whether these nomological propositions link only observable properties or both observable and non-observable properties, these law propositions will only be predictive in the first case or will have an explicative value in the second case. For example, propositions of thermodynamics are predictive propositions which help predict the evolution of a fluid with regard to the variation in one of its essential properties, but they are not able to explain this evolution. However, the kinetic theory of gases proposes, for the same facts, a construction that helps us to understand and explain this evolution.

A nomological proposition is an abstract system, i.e. a fiction, an abstraction of individuals' own cognitive processes. It is only accessible and sharable by means of systems of signs such as $L(x1, ..,, x_n, c_1, ..c_p)$, expressed in a natural or artificial language such as mathematics: $L(x_1, ..,, x_n, c_1, ..c_p)$ associated with semantic assumptions $\{SA_i(xi)\}$. The couple $L(x_1, ..,, x_n, c_1, ..c_p)$ associated with the assumptions $\{SA_i(xi)\}$ is a law statement. The way in which the nomological propositions are designated by law statements will be the focus of the next chapter.

Before discussing the truth value of a nomological assumption, one must first question its meaning. As described previously (section 3.4), the meaning of a proposition is formed, on the one hand, from the reference class of the proposition (e.g. gaseous matter, fluids and ultimate components of matter) and, on the other hand, from the intension of this proposition (its antecedents and consequences). The meaning of a proposition precedes its truth; this implies that only meaningful propositions may have a truth value. It is then possible to attach a truth value to a meaningful nomological proposition.

A nomological proposition only ever partially represents a factual law to the extent that such a proposition can only have an approximate truth value. The van der Waals gas law is only approximately true, but its truth value is greater than that of the ideal gas law.

For an assumption to be considered as a nomological proposition that is at least partially true, there must be a sufficient amount of empirical evidence confirming it. This evidence may result from direct or indirect observations or experiments.

Thus, the existence of Neptune, predicted by Le Verrier and introduced to confirm Newton's theory of gravity, required direct evidence such as the direct observation of Neptune by Johann Galle.

Similarly, the existence of the Higgs boson, introduced into the standard model of elementary particles to explain the diffraction of electroweak interactions, required even more indirect evidence insofar as its existence would be too short-lived to detect it and only the products of its disintegration would be observed. Detecting these products of disintegration, therefore, is a truth indication of the nomological proposition concerning the existence of Higgs boson.

3.6.2. *Applied sciences and technology*

The purpose of applied factual sciences and technology is to provide resources to act upon the real world, which are efficient since they are consistent with factually true law statements (approximately) and they are rationally based and rationally justified. This is what distinguishes factual applied sciences and technology, one the one hand, from fundamental sciences (as an autotelic activity) and, on the other hand, from techniques which resort to experience and routine without any other form of justification, as the philosopher Alain [CHA 60] expressed as follows: "What is the peculiarity of this technical thought? It is that it tries with hands instead of seeking through reflection".

Technology bears two aspects of fundamental and applied sciences: first, the scientific method that provides operative technological theories and, second, results of applied sciences that provide substantive technological theories.

3.6.3. *Operative technological rules*

Operative technological theories are applications of the scientific method to action, in other words, scientific theories for action. What first characterizes these operative technological theories is to present them as problem-solving processes, which go from stating the problem to be solved to the verification of a solution. Then, it resorts to mathematics and logic by means of theories such as theories of value, scheduling, games, linear programming or even queues. Finally, this is the lack of reference to particular factual sciences (physics, chemistry, biology, psychology and sociology). The operative technological theories apply to any type of objects without special consideration of manipulated objects.

Operations Research is the archetype of these operative technological theories. So, in [CHU 61, p. 13], Churchman *et al.* describe the different stages of an operations research project as a series of rules, including:

1) formulate the problem;

2) construct a mathematical model to represent the system concerned;

3) find a solution from the model;

4) test the model and the derived solution;

5) establish controls to the solution;

6) implement the solution: realization.

We can easily observe similarities between this method and the one recommended by the mathematician George Polya in his book *How to Solve It?* [POL 45] (in other words, how to formulate and solve a problem?), which is a heuristic method of mathematics based on problem-solving approach:

1) first principle: understanding the problem;

2) second principle: to formulate a problem-solving plan considering different solution strategies;

3) third principle: patiently and carefully execute the formulated plan;

4) fourth principle: reexamine and continue.

As we will see in the second part of this book, systems engineering methodologies fall completely into this category of operative technological theories.

3.6.4. *Substantive technological rules*

Substantive technological theories are derived from results of factual scientific theories (physics, chemistry, biology, psychology and sociology) into technological rules. In fact, what substantive technological theories do is they derive various technological rules from nomological propositions. Then, these rules can be used in different phases of action. Flight mechanics theories for airplanes or helicopters are derived from fluid mechanics as a fundamental science.

Engineers developing technological systems mainly base their action model on substantive technological rules.

These substantive technological rules are derived from nomological propositions such as the one in the following example, relating to the lift law of a wing (which is already included in applied science). This law is expressed by the formula: $F_Z = 1/2\,\rho V^2 S C_z$. From this, we can deduce the following technological rules:

1) To have the lift force F_z under control, we can modify the velocity V in proportions defined by the law $F_Z = 1/2\,\rho V^2 S C_z$.

2) To have the lift force F_z under control, we can modify the lift coefficient[11] C_z in proportions defined by the law $F_z = 1/2\,\rho V^2 S C_z$.

Regarding the nomological proposition from which they derive, these technological rules introduce an asymmetry between the intended effect (the property of the object that we seek to control) and the means used (the property of the object on which we will act) to meet this goal, while nomological propositions are symmetric (acausal). These technological rules become the components of reasoning process implemented in the design framework [MIC 06] .

Finally, it must be noted that if truth is the most important property of a scientific proposition, the efficiency of a technological rule prevails over its degree of truth. A technological rule T_1 derived from nomological proposition L_1 may be more efficient than a technological rule T_2 derived from nomological proposition L_2 whereas L_1 is less true than L_2 (i.e. L_1 provides less precise predictions than L_2). The most efficient technological systems do not necessarily use the most accurate scientific theories.

3.7. Knowledge and belief

To bring this chapter on knowledge systems to an end, we will describe the relationship between belief and knowledge. In fact, an engineer is constantly making decisions (about specifications, design, implementation and verification). Once they are made, he/she generally believes that these decisions are the best decisions or the less bad decisions possible in a given situation. If he/she wants to justify them, he/she refers to a certain amount of scientific and technological knowledge, confirmed for the first and efficient for the second.

---

11 As an example, high lift devices implement this technological rule.

Also, we recap the way in which Bunge [BUN 83] connects the concepts of knowledge and belief[12].

According to Bunge, if p is a nomological proposition, we can say that an individual s:

– believes p, if s knows p and holds p as true (or sufficiently true);

– is justified to believe p, if s knows p and if s knows a sufficient amount of evidences establishing that p is true (or sufficiently true);

– is justified to doubt p, if s knows p and if s knows that there are no bases to assign a definite truth value to p;

– is justified not to believe p, if s knows p and if s knows that p has been refuted.

Also, for a technological rule r, we can say that an individual s:

– believes that r is efficient, if s knows r and if it holds r as efficient (or sufficiently efficient);

– is justified to believe that r is efficient, if s knows a sufficient amount of evidences establishing that r is efficient (or sufficiently efficient);

– is justified to doubt that p is efficient, if s knows r and if s knows that there are no bases to assign r a definite efficiency value;

– is justified not to believe that r is efficient, if s knows r and if s knows that the efficiency of r has been refuted.

As a critical and reflective engineer (expression from [SCH 83]), he/she must therefore respond to the question of whether he/she is justified to believe to the substantive and

12 Usually hinged the opposite way, knowledge is, in the philosophical tradition, a true and justified belief (from a track of Plato's *Theaetetus* where knowledge is considered as a "right opinion provided with reason").

operative technological rules on which his/her design reasoning is based.

We will see that the validation process, as described in Chapter 8, is a process which gives us evidence to believe that correctly validated specifications are also as exact as possible, provided that it is conducted as rigorously as required.

Similarly, we will see that the verification process, as described in Chapter 9, is a process which gives us evidence to believe that correctly verified systems are also as consistent as possible with the specification, provided that it is conducted as rigorously as required.

4

Semiotic Systems and Models

4.1. Introduction

In the two previous chapters (Chapters 2 and 3), we considered technological systems and knowledge systems as two particular kinds of systems belonging to the general systems gender, as well as their specific properties. In this chapter, we are going to consider a third kind of systems: systems of signs that connect systems of the second kind to systems of the first kind. We will begin by defining what we mean by a system of signs, or (it is the same) a semiotic system, and we will introduce the denotation relationship that connects concrete systems and designation relationships that connect concrete systems to abstract systems. We will also introduce the concepts of signification and truth of systems of signs, inherited from that of meaning and truth of knowledge systems. We will then present what we mean by a law statement such as a system of signs, on the one hand, in relation with a nomological proposition of a knowledge system and, on the other hand, in relation with a factual law of concrete systems. Once we have presented these concepts, we will define what is meant by a model of a type of systems, using the concept of injective morphism. We will classify two types of models [BUN 73], object models and theoretical models, by specifying the type of morphism to which each refers. We will show how theoretical models pave the way to simulation, and we will conclude on the concepts of representativeness of models and expressiveness of languages.

4.2. Signs and systems of signs

First, a sign is a concrete object (drops of ink, a sequence of sounds, etc.) that allows a social group to denote another concrete object such as the word "airplane" which denotes any real airplane. It also helps designate the concept of "airplane" which is an abstract object. The concept of "airplane" can also represent real airplanes during cognitive activity.

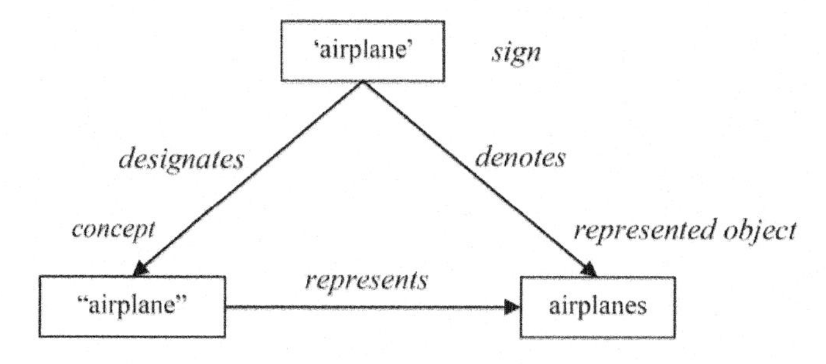

Figure 4.1. *Sign, concept and represented object*

Next, all signs do not necessarily denote real objects. Some denote nothing from the real world whereas they designate, in the same time, concepts. This is the case for the sign "π", for example, which designates a particular real number or even the word "unicorn" which does not denote anything real while it designates a poetic fiction. In other words, a system of signs is an abstract system that represents concrete systems or not. No concrete system is represented in the case for logical and basic mathematical theories.

If we take the sign "airplane", we can consider it to be a system of signs: its composition, the letters "a", "e", "i", "l", "n", "p" and "r" arranged into a certain order with repetitions. This order "*y follows x*" makes up the endo-structure of this system of signs. This system also has an environment of abstract objects such as the concept "airplane" it designates. The designation relationship belongs to the exo-structure of this system. This sign "airplane" has an emergent property, namely that of "having a signification", since its components do not have one. This signification is taken from the meaning of the concept to which it designates, i.e. the concept of an "airplane". The signification of a sign corresponds to the meaning of the designated concept. Generally speaking, the signification is a property of a system of signs inherited from the meaning of abstract systems they designate. Two signs have the same signification when they designate the same concept. This is why the sign *avion* in French has the same signification as the sign "airplane", or even the sign "plane" in English. These are synonyms.

Similarly, in the case of the Rosetta stone, the inscription includes a decree issued in Memphis in 196 BC on behalf of Pharaoh Ptolemy V. The decree affirms the royal cult of the new monarch. It is written in two languages (ancient Egyptian and ancient Greek) and three scripts: Egyptian in hieroglyphics, Egyptian in the Demotic script and the Greek alphabet.

The semiotic system, designating this decree, written using hieroglyphics (composition, endo and exo-structures), was successfully decoded by Champollion due to the coexistence of the other two semiotic systems, Greek and Demotic, which designate the same abstract system (namely, the proclamation of Ptolemy V as a god). The common signification between the systems of signs (Greek, Demotic and hieroglyphics) is inherited from the meaning of the designated abstract system, namely the proclamation of priests in Memphis.

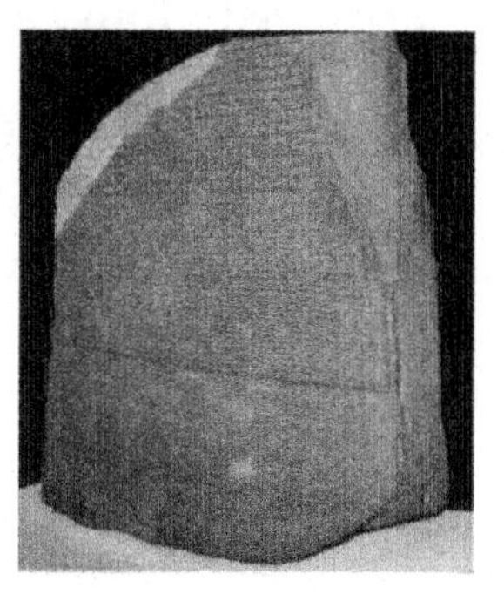

Figure 4.2. *Rosetta stone, signification and meaning*

If we now consider the system of letters "arilpane", in English, this sign does not designate any concept and does not denote any kind of concrete objects; in other words, it does not mean anything and does not correspond to any correct construction. This is not a sign; it is morphologically incorrect and semantically lacks signification. However, it shows some similarity with the sign "airplane", useful for orthographic correctors.

This is also the case for the Voynish manuscript. This is an illustrated book, written between 1408 and 1436, by an unknown author using an unknown alphabet. The nature of this unintelligible manuscript is still an enigma, whether a text for insiders or a joke; we can note that in the current state of things, it does not designate anything (it has no signification) and no longer denotes anything (it no longer has a reference).

Note that systems of signs may be put together to, in turn, form a semiotic system of a higher level; this is the case for sentences. Sentences are syntactically correct systems (endo-structure) of words, which designate abstract propositions and possibly denote facts (when the abstract propositions represent facts).

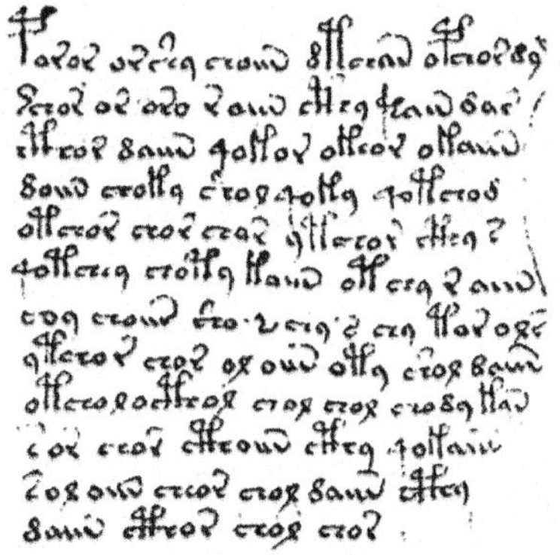

Figure 4.3. *Page extracted from Voynish manuscript*

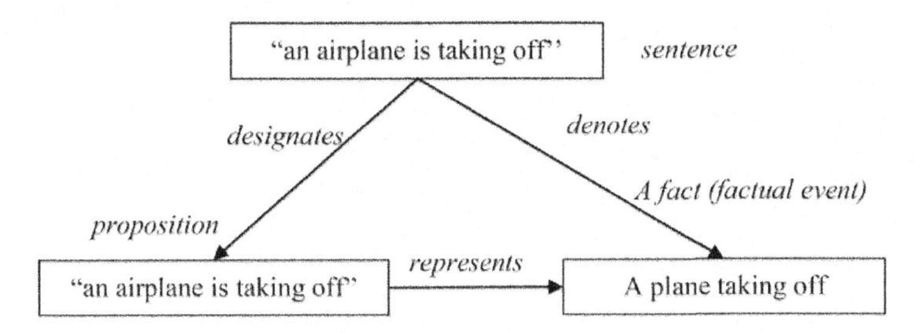

Figure 4.4. *Sentence, proposition and fact*

At this level, we can say that, on the one hand, sentences have taken signification from the meaning of propositions they designate, and on the other hand, they have an emergent property with regard to the previous level: they can have a truth value. A sentence is true if the proposition designated is true. Here, again, the endo-structure relates the morphology (here, syntax of the sentence), whereas its exo-structure relates the semantics.

Natural languages (written or spoken) are not the only type of semiotic systems that exist. We must also consider not only artificial languages, such as the language of mathematics, but also a plethora of systems of signs such as sheet music, as well as architectural designs, all imaginable types of mechanical, electrical, electronic, chemical and biological schematics, all imaginable types of diagrams, maps, chronographs and general physical or symbolic models.

Physical models, such as a scaled-down mock-up of an airplane inside a wind tunnel or an oil tanker inside a test basin, are also systems of signs, even if they are figurative objects. Note that there is also a gradation of semiotic systems from figurative (the shape of the signs has similarities with the shape of the represented object) to symbolic (with symbols, the shape of the sign no longer has any similarities with the represented object[1]) by passing through all the intermediate forms.

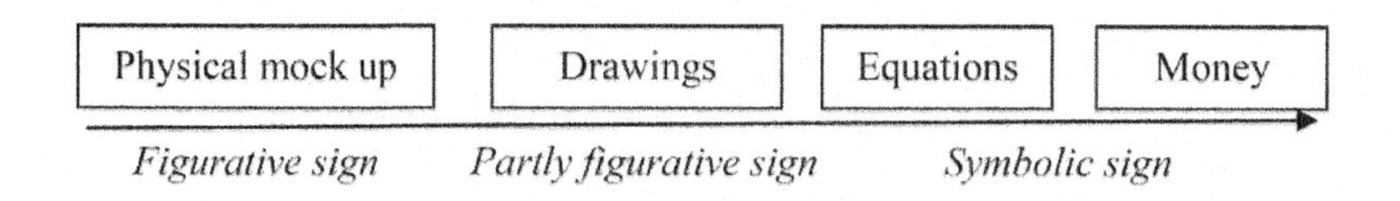

Figure 4.5. *Signs and symbols*

Therefore, a scaled-down model of an airplane at the counter of a travel agency is a figurative sign. Conversely, a cheque, its equivalent in paper money, coins, or its record in a ledger, and a bill of exchange are many symbols designating the same general right to acquire goods or services in an abstract system of monetary exchanges.

1 Symbols are purely conventional [SAU 00].

4.3. Nomological propositions and law statements

A nomological proposition (this concept was introduced in Chapter 3) is an abstract system, i.e. a fiction resulting from the abstraction of an individual's own cognitive processes. It is only accessible to other individuals and sharable among individuals by means of systems of signs, which we call law statements such as:

$$\left\{ \frac{L(x1, ..,, xn, c1, ..cp)}{\{SAi(xi), SAj(cj)\}} \right.$$

In this statement, the first term, $L(x_1, ..,, x_n, c_1, ..c_p)$, may be a mathematical formula linking together variables and constants, whereas the second term, {SAi(xi),SAj(cj)}, is made up of the semantic assumptions associated with this formula.

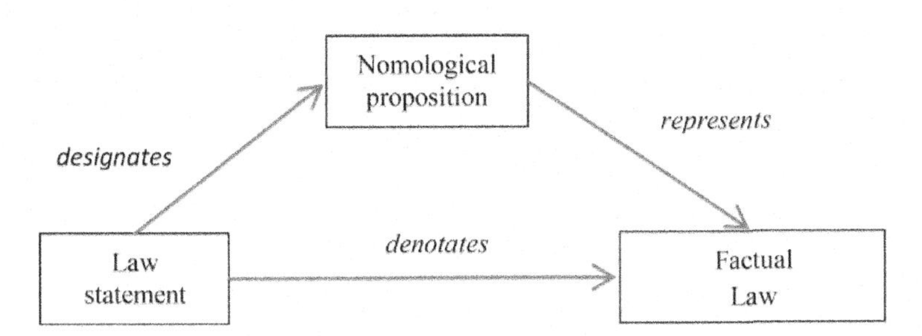

Figure 4.6. *Law statement, nomological proposition and factual law*

A formula, such as $F = 1/2\rho V^2 SC$, which links the signs F, V, S, C and ρ, has a signification and can be understood by anyone, when two conditions are fulfilled: (1) the formula is syntactically correct and (2) the signs F, V, S, C and ρ are signs designating concepts in the considered context. This association between signs in the formula and concepts of a domain, by means of semantic assumptions, completes the law statement. If F, V, S, C and ρ designate the concepts of lift force, velocity of the wing with regard to the surrounding fluid, characteristic wing area, etc., then this law statement (formula + semantic assumptions) designates a nomological proposition known as the lift law. The law statement, therefore, takes its signification from the meaning that the associated nomological statement bears.

Note that by changing the semantic assumptions associated with the same formula, we change the signification of this statement. In fact, if the signs F and C no longer designate the lift force, but instead the drag force, and so on, the statement changes its signification and designates the *drag law* instead of the *lift law*.

Moreover, since a nomological proposition does not accurately represent a factual law (this is only, at best, an approximately true representation), this is the same for the law statements that designate nomological propositions and denote factual laws. Thus, the van der Waals law statement is only approximately true. Similarly, the ideal gas law statement is also only approximately true. However, the truth value of the van der Waals law statement is greater than the truth of ideal gas law statement (the factual truth is not binary and comes in degrees).

4.4. Models, object models, theoretical models and simulation

We name *model* of an object type, a concrete object, from which we can, for example:

– produce copies of systems sharing the same type;

– explain the behavior (observable or not) of systems sharing the same type;

– predict the behavior of systems sharing the same type.

These operations are made possible due to a similarity between these systems (sharing the same type) and the model. We say that the model denotes this collection of systems or even that the model is a system representing the collection of represented systems sharing the same type.

Thus, the type definition description (TDD) is a model (concrete) of an aircraft type that designates an abstraction: the type definition (TD). The TD represents the collection of all produced aircraft[2] according to this definition.

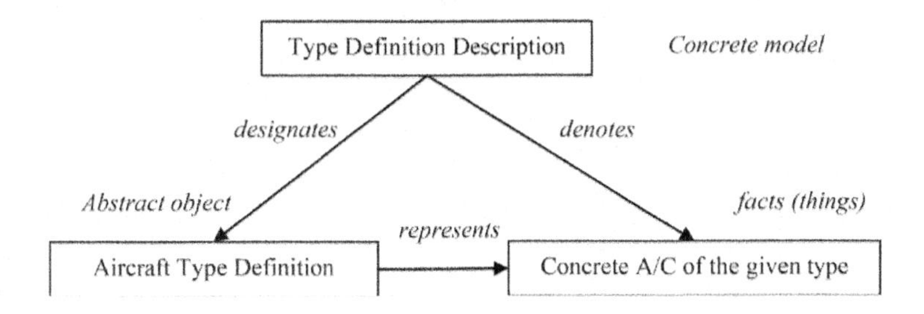

Figure 4.7. *Systems and models*

We name *object model* a particular type of model that establishes a mapping between the components, environment and structure of the model and the components, environment and structure of the represented systems. Thus, each component identified in the TDD by a part number (P/N) corresponds to a system, an equipment or a physical part (hardware or software) with P/N marks in each represented system. The location of these components in the represented system is provided by the TDD maps. Each connection diagram (electrical and hydraulic), identified in the TDD, defines an aspect of its structure, i.e. the way in

2 Or to be produced in the future.

which the different pieces of each represented system will be physically interconnected.

Formally speaking, the map between a model M and a represented system Σ by this model M is an injective morphism h of M in Σ, that is to say each component of the model has a concrete referent in the composition of a represented system (the reverse is not true). Each structural relationship of a model has a concrete referent in the structure of a represented system (the reverse is not true). Two distinct signs of the model denote distinct entities of the represented systems. The expression h(M), therefore, represents all components, the environment and the structure of Σ denoted by the object model M.

If A is a piece situated in zone E of the model (TDD), then the physical piece of the represented system, denoted by A, should be located in the physical zone of a represented system, denoted by E (and *vice versa*). If A and B are two components of the model, linked together by an electrical relationship in the model, then the physical pieces of equipment represented by A and B should be linked together electrically in a represented system (and *vice versa*). Or even, if the model of piece A is part of the model of B, then the physical piece A will be part of the physical equipment B in a represented system (and *vice versa*).

Behavioral models: however, the object model does not provide us with any information about the intended or actual behavior of the represented systems. In order to be able to predict or explain it, in the same way as for the object model, a link will be established between the evolution of states of the represented system and the evolution of the state variables of the model. This link can be formed in two ways: (1) concretely using physical mock-ups and prototypes or (2) theoretically, using theoretical models and simulation.

Physical models: so, a wooden mock-up of a wing or a blade inside a wind tunnel, a wooden mock-up of a cockpit and a prototype of a helicopter are figurative concrete models of a type of wing, blade, cockpit and a type of helicopter from which it is possible to assess the behaviors of a wing and a blade, the ergonomics of a flight deck and the performance of helicopters of a given type. This link is, therefore, a material similarity that links the common physical properties to the model and the represented systems: for example, the curvature and area of the blade model and that of the represented blade, the position and distance of two panels on the mock-up and in real cockpits.

Theoretical models: another method to establish a link is to design a theoretical model of the represented systems of a given type. A theoretical model is a projection on an object model of a general theory (mechanical, electrical, multiphysical and automatic) that denotes the laws linking different essential properties of represented systems together. A theoretical model T of a type of systems is therefore composed of an object model M and a specific theory S. The similarity is no longer a material similarity but a formal similarity. The semantic assumptions are a mapping between factual properties of the represented systems and the different signs of the model denoting these factual properties, whereas the factual laws that really link these properties are denoted by law statements.

For example, a specific theoretical model of the aerodynamic behavior of an aircraft can be derived from the general theorems of dynamic when they are projected on an object model of this aircraft. In the same way, a specific theoretical model of an aircraft equations of motion can be derived from the equations of fluid mechanics, when projected on an object model of an aircraft.

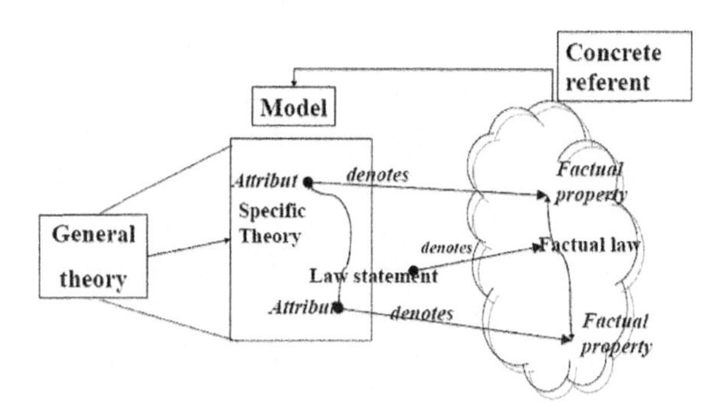

Figure 4.8. *Object model and theoretical model*

As with object models, we can define a mapping between the theoretical model T and the represented systems Σ, as a morphism h on the formally possible behaviors of the theoretical model, on the one hand, and the actually possible behaviors of the represented systems, on the other hand. The behavior B_{Σ} of systems of type Σ is the set of all actually possible trajectories in the space of states of Σ, whereas the behavior B_T of a theoretical model T is formed from the total formally possible trajectories in the space of states of T, provided that these formally possible trajectories have physical correspondent.

Based on this theoretical model T, it is possible to make predictions with regard to the behavior of the represented systems either by calculation or by simulation. Expressed in a simulation language, the law statements forming a specific theoretical model can be executed step-by-step using a simulator. The simulator provides the engineer with the evolution of parameters he/she wishes to observe according to the previously defined scenarios, such as assessing the fuel consumption during the performance of a particular loop flight, the stability margins available during a particular maneuver or even the effects of balance on the stability and/or consumption of fuel.

4.5. Representativeness of models and the expressiveness of languages

All models are not equivalent; some models may be more representative than other models. Similarly, all languages used to define models are not equivalent. Some languages may be more expressive than others.

4.5.1. *Representativeness of models*

Object model representativeness: first, let us consider object models M of one type of systems Σ. Assuming that M_1 and M_2 are two object models of the same type of systems Σ, there are two morphisms h_1 of M_1 in Σ and h_2 of M_2 in Σ. If $h_1(M_1)$ is the image of M1 in Σ and $h_2(M_2)$ is the image of M_2 in Σ, then two situations can occur:

1) $h_1(M_1) \subset h_2(M_2)$ or $h_2(M_2) \subset h_1(M_1)$: in this case, one of the two models denotes one part of the represented systems that is included in that denoted by the other model. In this case, we can say that the first model is less representative than the second model. For example, if $h_1(M_1) \subset h_2(M_2)$, then M_1 is less representative of the represented systems than M_2.

2) $h_1(M_1)\ \Delta\ h_2(M_2) \neq \emptyset$: in this case, both models denote distinct aspects of the represented systems and hence they are not comparable.

An object model of a type of systems Σ is more representative than another object model if it provides more details about the composition, environment and structure of the represented systems. For example, a multilevel object model is more representative than an object model with only one level that does not express the whole-part relationships. Similarly, an object model that expresses neither the component attributes nor the relationships and the entities of the environment is less representative than a model that does so.

Theoretical model representativeness: now, let us consider two theoretical models T_1 and T_2 of a type of systems Σ. If $B(\Sigma)$ represents all the actually possible behaviors of Σ, then $B(T_1)$ and $B(T_2)$ represent all the formally possible behaviors of models T_1 and T_2. There is an injective morphism h_1 of $B(T_1)$ in $B(\Sigma)$ (respectively, h_2 of $B(T_2)$ in $B(\Sigma)$), which for each formal behavior b_1 of $B(T_1)$ matches a behavior $h_1(b_1)= b$ of $B(\Sigma)$.

Let b_1 (respectively, b_2) be a behavior of the theoretical model T_1 (respectively T_2) denoting the same actually possible behavior b of Σ, then b_1 is more representative of b than b_2 if b_1 is more accurate than b_2 (i.e. the distance from b_1 to b[3] is shorter than the distance from b_2 to b).

Now, if $h_1(B(T_1))$ and $h_2(B(T_2))$ denote the images of $B(T_1)$ and $B(T_2)$ in $B(\Sigma)$, then:

1) T_1 is more representative than T_2 of Σ if (1) $h_2(B(T_2)) \subset h_1(B(T_1))$ and (2) for all b of $B(\Sigma)$, b_1 denoting b in $B(T_1)$, b_2 denoting b in $B(T_2)$, b_1 is more representative than b_2 of b.

2) T_2 is more representative than T_1 of Σ if (1) $h_1(B(T_1)) \subset h_2(B(T_2))$ and (2) for all b of $B(\Sigma)$, b_1 denoting b in $B(T_1)$, b_2 denoting b in $B(T_2)$, b_2 is more representative than b_1 of b.

3) In other cases, both models T_1 and T_2 denote either different aspects of the behavior of the represented systems Σ or the same aspect of the behavior of the represented systems Σ but with globally incomparable representativenesses: T_1 and T_2 representativenesses are not globally comparable.

In all cases, we can say that the theoretical model T supported by an object model M is more representative than the object model M itself.

3 More precisely, a set of empirical data relating to b.

Whatever the model of a type of systems Σ can be, it always has a limited representativeness. In a model, some aspects of reality are inevitably overlooked. This is why the morphism associating a model with the systems it denotes cannot be an "isomorphism".

4.5.2. *Expressiveness of a language*

Now, if we examine the languages used to build object or theoretical models, we cannot exclude any, whether natural or artificial, literal or graphical. We can produce a model in any language, but the result will strongly depend on the language used.

Some of these languages have a limited expressiveness, that is to say, they can only produce one type of models such as the language of fault trees. This is a graphical artificial language that graphically denotes the possible combinations of events leading to a predefined undesirable event and supports the associated probability calculations. This is very useful when carrying out safety analyses, but it is of no use in many other situations.

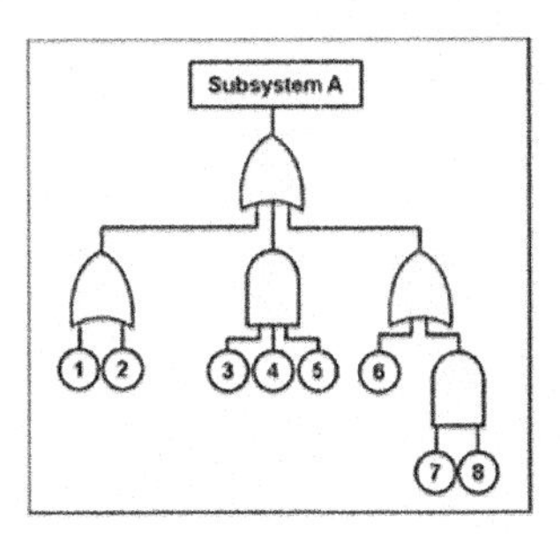

Figure 4.9. *Fault tree representation*

We can show that this type of fault representation can have equivalent representations such as reliability block diagrams with the same representativeness.

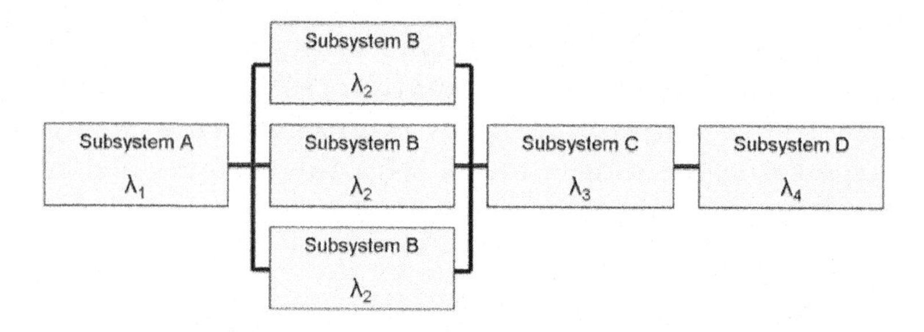

Figure 4.10. *Reliability block diagram*

The language of fault trees and that of the reliability block diagram have the same expressiveness [MOD 10].

Let us compare artificial languages, such as VHDL-AMS [IEE 07] and VHDL [IEE 08]. These are two literal languages; the first is more expressive than the second in the sense that it allows the production of object models and theoretical models of discrete, continuous or hybrid systems, whereas the second only allows the production of object models and theoretical models of discrete systems.

Now, let us compare the artificial languages SA-RT [WAR 86] and SysML [OMG 12]. These are two graphical languages; the first is less expressive than the second in the sense that the first allows the production of less representative models than those produced using the second, but both have a limited capacity to produce theoretical models.

Finally, natural languages are those whose expressiveness is the greatest. This is the reason why object models or theoritical models are often annotated with additional comments in the form of semantic assumptions which the modeler adds to clarify his/her intention.

Nevertheless, natural languages do not guarantee the production of homomorphic models of the systems they denote. On the contrary, due to the abundance of figures of speech, they vary the way in which an object, an event or any other feature or behavior of a system can be denoted. These figures of speech sometimes appear as additional text, sometimes by deletion or removal, or even by rearrangement or substitution. This stylistic abundance at the engineer's disposal poorly accommodates linguistic and cultural differences within a work community and is in direct conflict with the requirement to construct homomorphic models of the systems they represent.

This is why the engineering approach to model-based systems tries whenever possible not to use natural languages and instead use artificial languages adapted to producing homomorphic models of the represented systems.

PART 2

Methods

5

Engineering Processes

5.1. Introduction

If we refer to the state of the art and to professional standards, different processes will be applied by a project team whose aim is to develop a technological system.

The name, designation and the way in which these processes are organized depend on the methodological context in which they are found. For example, engineering processes considered in the recommendation ARP4754A[1] and standard EIA 632 [ANS 03] are not strictly identical for exactly this reason. Therefore, ARP4754A, recommended for the development of passenger transport aircraft and their subsystems, pays special attention to the safety assessment process, and to its dependencies and interactions with other aircraft development processes, although this process is not explicitly mentioned in other standards such as EIA632 or even ISO15288 [ISO 05].

Similarly, the concepts around which these development processes are organized also depend on the methodological

1 EUROCAE ED79A/SAE ARP4754A, "Guidelines for development of civil aircraft and systems", EUROCAE, France, December 2010.

context in which they are found. This is not surprising given the fact that these methodological frameworks involve technological knowledge systems and that the meaning of a concept is an emergent property of such an abstract system. For example, the term "validation" in the recommendation ARP4754A and standard ISO15288 refers to diametrically opposed concepts. In the first case, the validation is a process that involves ensuring that the system requirements are sufficiently correct and complete and that the assumptions made about the environment of the system are also sufficiently correct and complete. In the second case, ISO15288, the validation is a process that allows us to confirm, via a certain number of tests, that the requirements are satisfactory ("confirmation, through the provision of objective evidence, that the requirements for a specific intended use or application have been fulfilled", ISO15288, section 4.23, p. 5).

In other words, development standards, such as EIA632, ISO15288 or even ARP4754A, are knowledge systems. These are technological knowledge systems in the sense that they state rules for technological process performance. Moreover, these rules are not substantial as those described in Chapter 3, as they are not derived from factual scientific laws. These engineering rules are operative rules (see section 3.6.3) defined to produce an effect, and they are based on scientific method as we can also see in works on operations research.

Moreover, in the infancy of systems engineering, Arthur D. Hall [HAL 62], considered as the father of this discipline, highlighted the current similarities between operations research and this new discipline which he called systems engineering:

> The intellectual processes comprising the patterns of operations research and systems engineering show more similarities than

> differences. The phases of operations research listed by Churchman *et al.* are as follows:
>
> 1) formulating the problem;
>
> 2) constructing a mathematical model;
>
> 3) deriving a solution from the model;
>
> 4) testing the model and the solution derived from it;
>
> 5) establishing controls over the solution;
>
> 6) putting the solution to work: implementation.
>
> The similarities with the phases of systems engineering are obvious. This is not surprising since both are elaboration of modern scientific method applied to different ends.

These standards have many similarities, but they are not necessarily the same. They may be more or less coherent, more or less complete, which is why we will find our own way through the different recommendations they provide. This is how we will proceed in the following chapters.

5.2. Systems engineering process

5.2.1. *General framework*

In this section, we introduce the general framework that is referred to in this book, namely the eight processes defined by ARP4754A (see Figure 5.1).

The planning process forms the basis of systems engineering processes according to ARP4754A. They anticipate that, for each process identified, a plan is established and followed throughout the whole development or redevelopment process. These plans will be followed; the variances shall be recorded and the plans shall eventually be revised.

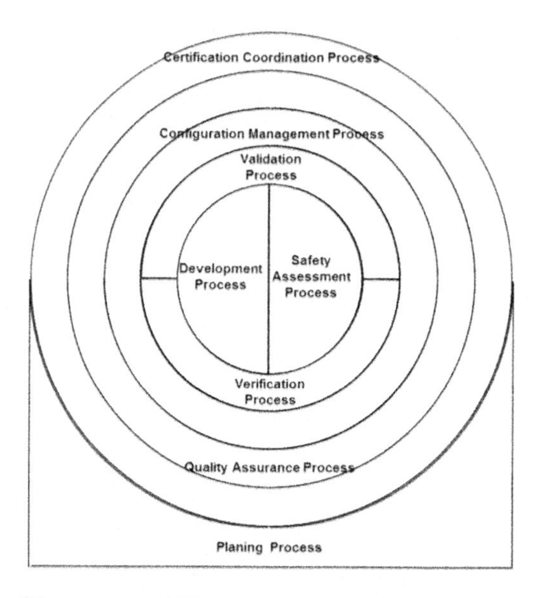

Figure 5.1. *ARP4754A engineering processes*

The seven other processes are organized and displayed according to an onion skin model: (1) development and (2) safety assessment are in the center and are presented as two halves of a disk to illustrate an objective of perfect integration. Around these two core processes, the following five processes are evolved: (3) the validation process of requirements and assumptions, which is located upstream and (4) the verification process of implementation that is downstream. In this onion skin description, these two processes are then successively followed by (5) the configuration management process for the system under development or redevelopment, (6) the quality assurance process (more specifically, process assurance) and (7) the certification process that organizes the compliance demonstration of the installed system with the airworthiness regulations.

5.2.2. *Design process*

Nevertheless, the development process described in ARP4754A is only a fairly general process (ED79A/ ARP4754A, Chapter 4 page 16), and we consider that this process needs to be described in more detail. We aim to do this by quite simply referring to the standard ANSI/EIA632 that proposes an iterative design process for a system by working through the activities at each stage of the design process as detailed below.

A technological system is composed of an assembly of building blocks, as shown in Figure 5.2.

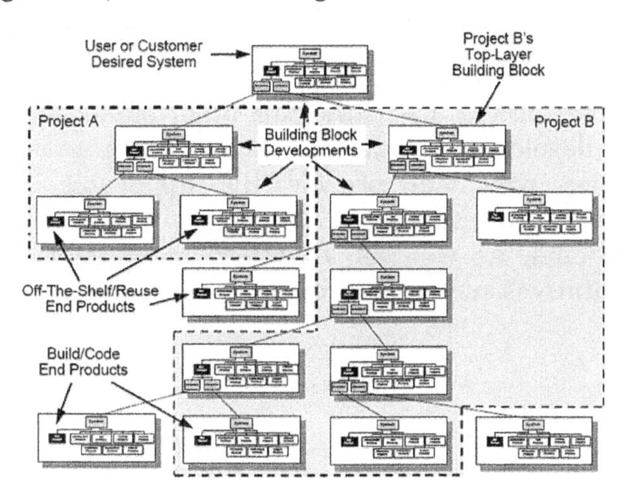

Figure 5.2. *EIA632 system breakdown structure*

For each of these building blocks, the same design process is applied, except for the terminal blocks.

The terminal blocks are either off the shelf or reused, such as equipment issued with a letter of approval TSO[2] or

2 TSO: Technical Standard Order. A TSO is a minimum performance standard for specified materials, parts and appliances used on civil aircraft issued by FAA.

ETSO[3], or blocks directly manufactured or coded (for software).

The design process for each block involves two subprocesses:

1) a subprocess defining the requirements assigned to the considered building block;

2) a subprocess defining a solution.

The first subprocess involves the following tasks:

– gathering the acquirer requirements: either the requirements of the system acquirer or user, or the requirements specified at a higher level and assigned to the considered block when in an intermediate or terminal level;

– gathering requirements from other stakeholders of the system development project, such as the airworthiness authorities who will provide the applicable regulation requirements, the safety analysts who will provide the safety requirements, as well as all other engineering discipline representatives such as maintenance engineers.

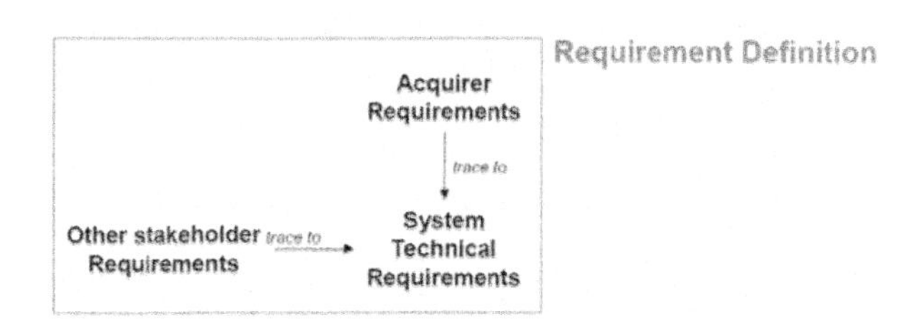

Figure 5.3. *EIA632 building block requirement definition subprocess*

The second subprocess involves the following tasks:

– elaborating a logical solution to the considered block: the standard does not decide the nature of this logical solution

3 ETSO: European Technical Standard Order issued by EASA.

but it can be, for example, composed of a network of processes linked together by flows of matter (e.g. fuel), energy (e.g. electric or hydraulic) or data flows and linked to one another by signals (event flow). We call the result of this logical solution the behavioral architecture or behavioral design[4] of the block. This behavioral design describes the intended behavior of the block. Of course, there may be several possible behavioral architectures of a building block. Selecting the preferred logical solution is an assessment and a decision-making issue between the advantages and disadvantages of the different solutions considered [MIC 06];

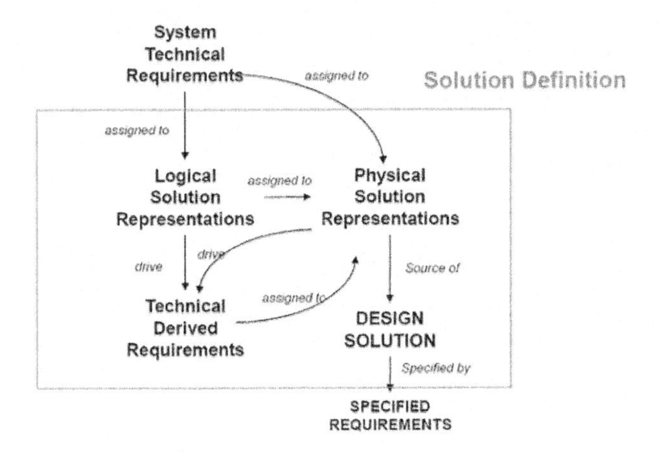

Figure 5.4. *EIA632 building block solution definition subprocess*

– elaborating a physical solution of a considered block: the standard does not decide the nature of this physical solution but it can be, for example, composed of a network of processors, diverse actuators, pumps, pistons and electric motors linked together by circuits (e.g. fuel, hydraulic, electric circuits and data bus). We call the result of this physical solution the physical architecture or physical design

4 We prefer the term "behavioral" to that of "functional" (see Chapter 6).

of the block. This physical design of the block describes the structure of the block or, in other words, the physical architecture of a solution. Of course, there may be several possible physical architectures of a building block. Selecting the preferred physical solution is also a decision-making issue, as mentioned above.

Inputs and outputs: the inputs of these two subprocesses are the technical requirements of the system and the outputs are the specific requirements of the design solution.

Assigning the technical requirements of a system: the upstream requirements are assigned according to their nature (structural or behavioral requirement) to the logical solution (for example, a functional, performance, time response and accuracy requirement) or directly to the physical solution (for example, weight requirement, size and geographical location). This is the requirement assignment process to the logical and physical solutions.

Technical requirement derivation process of a system: once assigned to a logical or physical solution, these requirements are derived and assigned to different elements of a logical (respectively, physical) solution. As shown in Figure 5.4, the logical and physical solutions determine the requirement derivation process. This means that the derived requirements depend not only on the requirements from which they derive, but also on the solution (logical or physical). In other words, there is no derived requirement that does not depend on the design solution in which this requirement was formed, except perhaps when it comes to poor requirement engineering practices.

Assigning elements of a logical solution to those of a physical solution: finally, elements of a logical solution, accompanied by derived requirements, are assigned to the physical solution. It results in a design solution that describes all components and physical links of the solution, elements of a logical solution (functional) that have been assigned to each one, as well as the behavioral (logical) and structural (physical) requirements assigned to them. The requirements assigned to each element of a design solution are specified requirements.

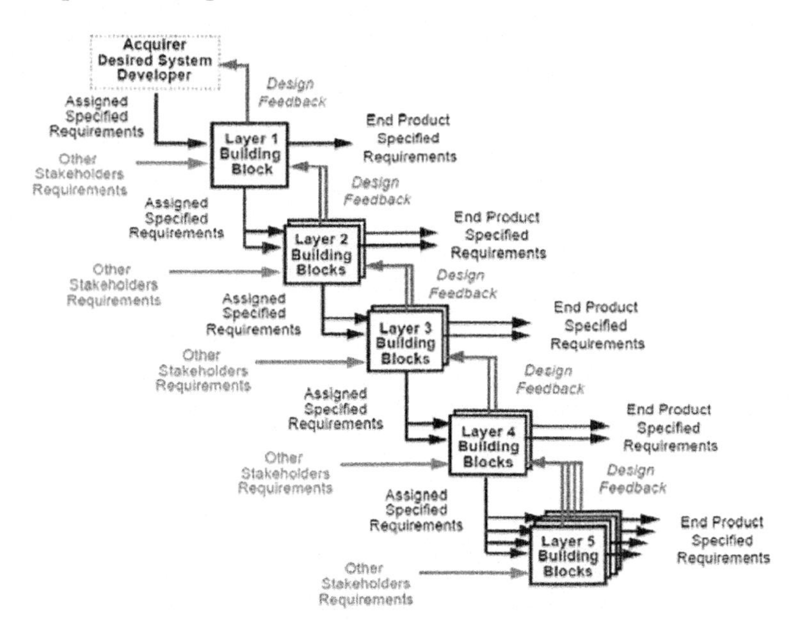

Figure 5.5. *EIA632 building block design processes*

In the case of a multilevel system, the process begins at the highest level of the system. It is then repeated for each building block for which, at the input of each block at level n, there are acquirer requirements and other stakeholder requirements, whereas at the output, there are requirements specific to the components forming the design solution of the block at level n. The process terminates at the lowest level blocks. These blocks, at the lowest level, are either blocks available from an equipment manufacturer, such as an equipment manufacturer under TSO/ETSO, or blocks that will be developed or modified according to a specification by an external or internal supplier, or even, directly codable blocks (software) or directly manufactured without other design activity.

5.2.3. *Safety assessment process*

These processes involve the different analyses such as:

– functional hazard assessment (FHA);

– preliminary safety assessment (PASA and PSSA);

– common cause analysis (CCA);

– safety assessment (ASA and SSA).

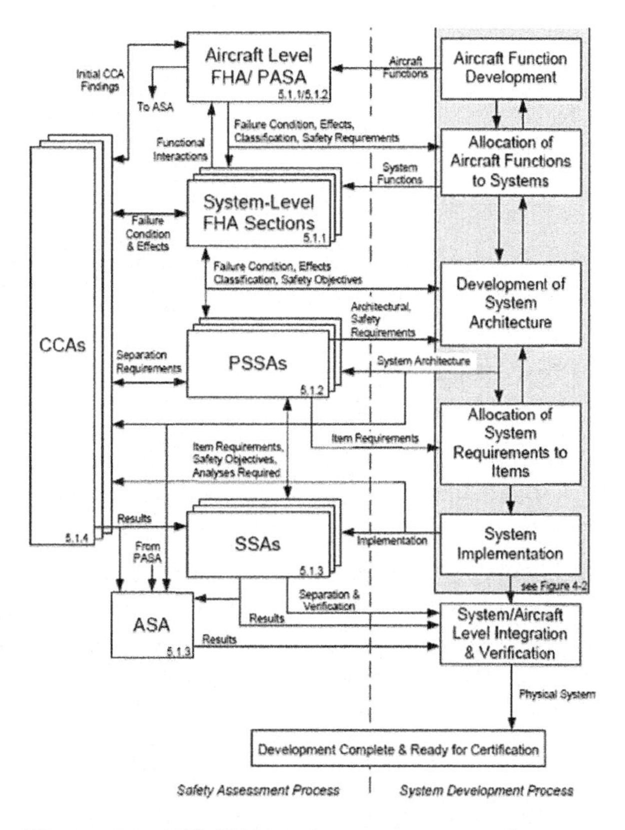

Figure 5.6. *ARP4754A safety assessment and system development process integration*

ARP4754A highlights the necessity of carrying out the developmental and safety assessment tasks in an integrated manner to keep the applicable requirements consistent as well as with the design choices and safety analyses to which they refer. This is shown in Figure 5.6.

To meet the integration requirement of safety assessments activities and those of requirement definition and solution definition, we propose an extension of the framework provided by EIA 632 by introducing safety assessment tasks at each design block level.

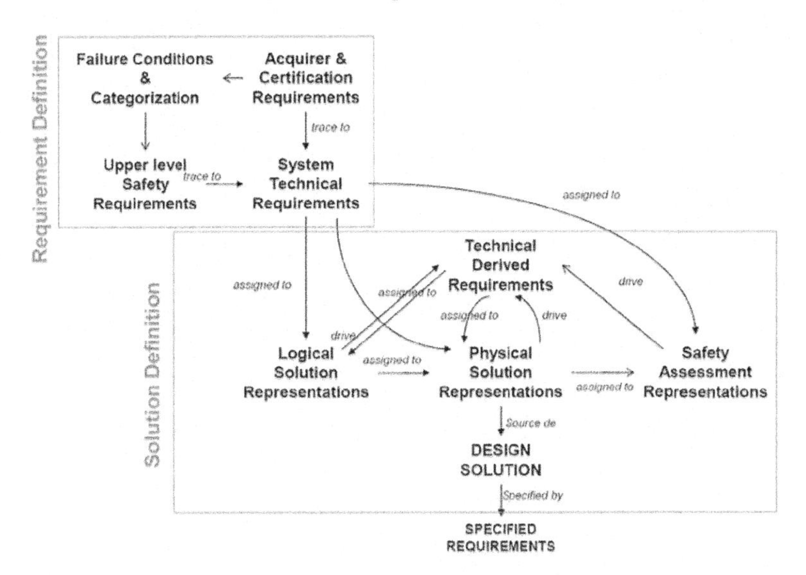

Figure 5.7. *EIA 632 design process extended to safety assessment topics*

As far as the requirement definition process associated with the considered block is concerned, we propose adding a safety requirement collection activity, conducted as part of the system's technical requirement definition, and based on functional hazard assessment (FHA) carried out at the level of the building block.

However, for the solution definition process associated with a given block, we propose formulating a solution for safety assessment (such as dependence diagrams and fault trees) of the considered block. The elements of a physical solution include the solution framework for safety assessment, i.e. the associated dependence diagram (or reliability block diagram). The safety requirements belonging to the technical requirements of a system are generally assigned to these representations and may be derived and assigned to different elements of a physical solution. The design solution as well as the requirements specific to each element of a design solution may have additional safety requirements assigned to them.

5.2.4. *Requirement and assumption validation*

ARP4754A recommends carrying out a validation of the requirements and assumptions at each decomposition level of a system by preferably beginning with the highest level; we will describe requirements and assumptions in more detail in Chapter 6.

The purpose of this validation process is to guarantee that the technical requirements of the system, as well as the specified requirements, are as correct and complete as possible, at each decomposition level of the system, and therefore the system under development is the desired system and the development equipment is appropriate for the purpose. Similarly, all assumptions regarding the definition of these requirements shall also be validated.

"The requirements shall be sufficiently complete" means that overlooked requirements should not make the system unacceptable to the stakeholders "(the client, user, authorities etc.)". They shall also be sufficiently correct, that is to say they shall comply with the requirements to satisfy the desired type of system.

The validation process of requirements and assumptions of a system is a long and costly process. However, it is less costly than developing a type of systems inappropriate for the intended service. At the highest level, the validation of requirements and assumptions is more of a marketing, financial and regulatory problem; therefore, it has more to do with economic and social engineering than the engineering of technological systems. However, once this first stage achieved, the validation of requirements and hypotheses is handed over to technological engineering. It is, therefore, more about showing that the derivation of system requirements has been correctly and completely carried out and that the derived requirements obtained correctly and completely cover the original requirements. This is what we propose to do in the validation chapter of this book.

Finally, ARP4754A recommends that the validation effort should be proportionate to the severity of failures of the developed systems and the validation tools, such as simulation, analysis, tests and expert judgments, are used gradually, and possibly in combination, according to the severity of the failures. Some of these tools, such as simulation, may be used early in the development cycle, whereas others, such as flight testing, are likely to be used later. Although they are complementary to each other, it is clear that creating a validation strategy, primarily concerning time-consuming methods, such as flight tests, is both risky and costly. This is the reason why we propose, in Chapter 8, a strategy that puts emphasis on simulation as an early validation means.

5.2.5. *Verification of the implementation regarding requirements*

ARP4754A recommends proceeding in a complimentary manner to requirement validation, a verification of the implementation facing the requirements, at each level of system decomposition.

The purpose of this verification process is to guarantee that the physical realization of each block conforms to the assigned requirements. With each level of requirements having been validated, we can also expect that the system obtained will conform to initial expectations.

The different stages of realization include design solutions of each block, the physical realization of each elementary block and the gradual integration of blocks, until the whole system is obtained. ARP4754A recommends that the verification focuses on all physical products. However, we consider that it should also involve products of design (design descriptions and design models).

Finally, ARP4754A recommends that the verification effort should be proportionate to the severity of failures of the developed systems and the verification means, such as simulation, analyses, inspections and expert judgments, should be used gradually, and possibly in combination, according to the severity of the possible failures. Some of these means, such as simulation, may be used early in the development cycle, whereas others, such as flight testing, are likely to be used later. Although they are complementary to each other, it is clear that creating a verification strategy, primarily concerning time-consuming methods, such as flight tests, is both risky and costly. This is the reason why we propose, in Chapter 9, a strategy that covers each stage of the implementation, including the design process.

5.2.6. *Managing configurations*

Although this topic is not widely discussed in this book, it is not a problem regarding the context of our approach;

ARP4754A recommends implementing a configuration management process that assures controlled development. The configuration of definition products, the products implemented, the means of validation, implementation and verification shall be established and controlled. The corresponding data shall be archived and retrievable. Defects in the manufactured products shall be recorded as problem reports, the causes of these defects shall be located and errors in the product definition at the origin of these defects shall also be located and recorded. Changes may have several causes: (1) errors, (2) not related to errors but eventually to a change in the need or the context or even (3) a progressive development strategy by means of versions. These modifications shall undergo a process that allows the product configuration to evolve in a mastered way.

5.2.7. *Process (quality) assurance, certification and coordination with authorities*

ARP4754A recommends implementing a process (quality) assurance process that ensures conformity of all processes actually implemented in a system development project in relation to plans defined in the planning process (each process shall follow the plan assigned to it) in order to identify possible errors, and to follow actions to reduce those errors. At the same time, proof of its own activities shall be gathered.

Finally, ARP4754A recommends implementing a certification coordination process with the authorities, which allows the certification basis to be determined, and to together agree on the method of demonstrating the compliance of the system installed in an aircraft with the certification basis and to provide the authorities with evidence of this compliance to the established and agreed plan.

In this book, although we do not deal with these points, it is certain that what we do propose has a positive impact on certification processes and, in particular, on the demonstration of compliance with general requirements such as FAR25.1309, FAR27.1309 and FAR29.1309 in USA or CS25.1309, CS27.1309 and CS29.1309 in Europe.

6

Determining Requirements and Specification Models

6.1. Introduction

As pointed out in the previous chapter, the composition of a technological system is made up of building blocks that its endo-structure assembles, as shown in Figure 5.2. For each building block, the same design process is applied, except for the terminal blocks, to which a specific process is applied.

The design process of each block consists of two subprocesses:

1) a subprocess defining the requirements assigned to the considered block;

2) a subprocess defining a solution.

The first of subprocess involves the following tasks:

– gathering acquirer requirements: either, at the top level, the client requirements or those of the system user, or, at an intermediate or terminal level, the requirements derived from the upper level and assigned to the considered block;

– gathering requirements from other stakeholders of the system development project, such as certification authorities who supply the applicable regulatory requirements, the safety analysts who provide the safety requirements, as well as all other engineering discipline representatives such as maintenance engineers.

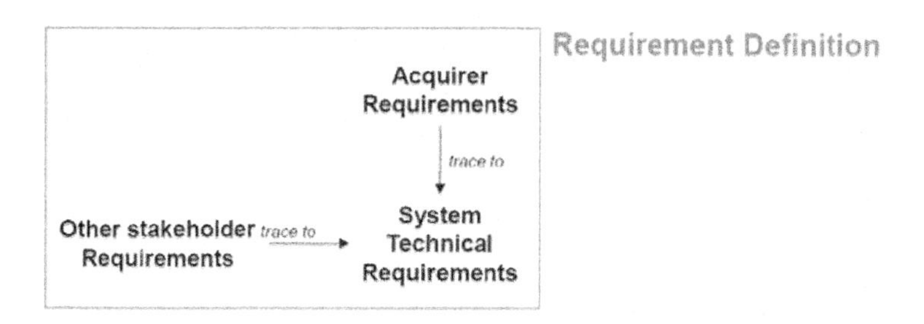

Figure 6.1. *EIA632 building block requirement definition subprocess*

The second subprocess, called the design solution process, will be dealt with in the next chapter, whereas the current chapter is dedicated to the determination of requirements, that is to say, establishing the specification of requirements assigned to a system.

In the property-model methodology approach, which is based on models, such requirement specifications are called specification models. We will associate with the considered type of systems, and its subsystems, specification models, as already recommended in the EIA 632 standard.

Instead of specification documents written using tools such as Microsoft Word, IBM DOORS and others, the specificity of property-model methodology (PMM) lies in specification models that are compiled and executable in simulation environments, when they are subjected to simulation scenarios.

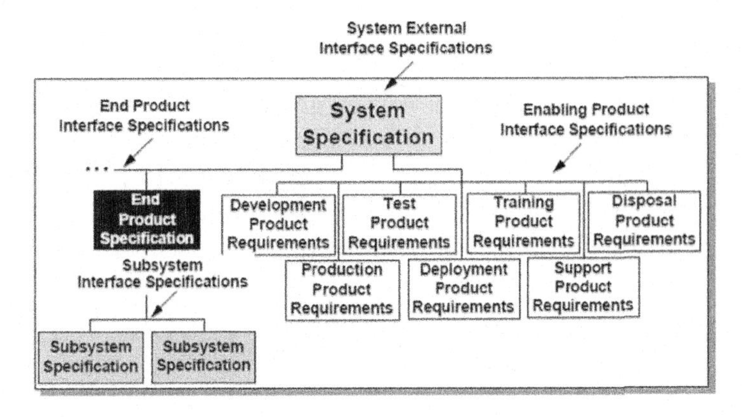

Figure 6.2. *EIA632 system specification architecture*

We will introduce here the concepts of systems specification, overspecification, underspecification and exact specification. After considerations of the subjective character of text-based requirements (TBRs) and text-based specification, we will introduce the concept of property-based requirement (PBR). On this concept, we will develop a theory of these PBRs, which is consistent with the model-based systems engineering (MBSE) paradigm. In particular, we will define a partial-order relation (comparison of requirements) and a conjunction operator among requirements, based on Bunge's theory of properties[1] (see section 1.5.2). We will then present an interpretation process that allows us to translate TBR into one or more PBRs. We will give examples of this interpretation process of TBRs coming from the regulations. We will conclude this chapter with a discussion on the relationship between specification models and concurrent assertions of simulation languages.

In Chapter 7 (section 7.2), we will introduce the concept of requirement derivation, whereas in later chapters, we will

1 The initial spark of the PBR theory partially comes from the chapter 3 of "System Requirements Analysis", written by Jeffery O Grady, and on an intellectual inheritance from B language, invented by J.R Abrial.

deduce, from them, two theorems of practical importance: the first (mentioned in section 9.3.5) which we call the contract theorem assures the conformity of a system, under certain conditions, with its specification model, whereas the second theorem (mentioned in section 10.3.7) concerns the impact of additional requirements introduced at a subsystem level, on the feasibility and safety of a type of systems.

6.2. Specifications and requirements

Here, we consider a collection of concrete objects composing a reference collection[2] K, for example, the collection of geostationary artificial satellites or the collection of transport helicopters or even the collection of nuclear power plant units.

Specifying a collection E of entities of type K, to be developed, means defining well-formed requirements $\{Req_i\}_{1 \leq i \leq n}$, such that any entity e belongs to E if and only if e simultaneously satisfies all the requirements Req_i of $\{'\}_{1 \leq i \leq n}$.

This can be formalized as follows: $E = SAT_K(\{Req_i\}_{1 \leq i \leq n})$ where $SAT_K(\{Req_i\}_{1 \leq i \leq n}$ denotes the collection of entities of type K that shall satisfy the requirement specification $\{Req_i\}_{1 \leq i \leq n}$

Now let us consider a collection E of entities of type K and a specification $S = \{Req_i\}_{1 \leq i \leq n}$:

– We say that S is an *exact* specification of E if $SAT_K(S) = E$, i.e. if the specification S accurately covers the desired collection E.

2 In this section, we use the term "collection" rather than the term "set" to highlight the dynamic character of a collection: a collection may be empty at a given moment such as the collection of aircraft corresponding to an ongoing definition, then to a variable number of copies when the first specimens, corresponding to a given definition, exit the assembly lines and put into operation.

– However, S is an overspecification of E if $SAT_K(S) \subset E$ and $SAT_K(S) \neq E$; in other words, if some entities of E are unduly excluded by the specification S.

– Similarly, S is an underspecification of E if $E \subset SAT_K(S)$ and $SAT_K(S) \neq E$, i.e. if S is not selective enough and does not exclude entities that do not belong to E and should not be in $SAT_K(S)$.

– There are specifications that fail from all points of view, if $E \not\subset SAT_K(S) \not\subset E$, i.e. if S is too selective in some aspects by excluding entities that belong to E and, not selective enough by others and does not exclude entities that do not belong to E.

– Finally, there may be unfeasible specifications S. These specifications are characterized by $SAT_K(S) = \emptyset$, i.e. there is no concrete object of type K that responds to specification S. In this case, we can say that S is unfeasible, or the requirements of S are conflicted or contradictory. We can note that this contradiction may be either absolute or relative. It is absolute when two or more requirements go against the "laws of nature", that is to say, an unjustified objection to a true knowledge item. It is relative when the state of scientific and technological knowledge, in a given society, leads to impossibility at a given moment, whereas this contradiction may be subsequently resolved due to discoveries or inventions.

It is incredibly easy to produce inaccurate specifications; practice shows us this repeatedly. This is why a specification shall be validated, as already mentioned with standards such as EIA 632 or an aeronautical recommendation such as ARP4754A. The validation process discussed in Chapter 8 is a process that aims to produce specifications as exact as possible.

Clearly, there may be many ways to specify a collection E of entities to be developed. We then obtain: E =

$SAT_K(\{Req_i\}_{1 \leq i \leq n}) = SAT_K(\{Req_j\}_{1 \leq j \leq m})$, and the specifications $\{Req_j\}_{1 \leq j \leq n}$ and $(\{Req_i\}_{1 \leq i \leq m})$ are then technically equivalent. However, these technically equivalent specifications may not be equivalent from an economic point of view. Some of them may be less costly to develop than others.

Intuitively, we can think that even if they are technically equivalent, the specifications with a limited number of requirements, which are comprehensible and easy to implement and test, are more desirable than specifications with many requirements, which are difficult to understand, implement and test. The second type of specifications are more complicated and costly than that of the first type.

The purpose of determining the requirements of a collection E of entities is to define an exact specification S of E that is as simple as possible. This is our objective in this chapter.

6.3. Text-based requirements and subjectivity

Usually, a requirement is defined as a statement that expresses either an obligation or a prohibition. One of the most consistent definitions is provided by the IEEE 1220 standard [IEE 05] which specifies that a requirement is "a statement that identifies a product or process operational, functional, or design characteristic or constraint, which is unambiguous, testable or measurable, and necessary for product or process acceptability (by consumers or internal quality assurance guidelines)". Other standards (e.g. EIA 632, ISO15288 and ARP4754A) are much more vague with regard to the definition of "requirement". In particular, ISO15288 does not define the term and ARP4754A is a good example of a window dressing (*trompe l'oeil*) definition: to show this point, it is sufficient to compare the definition of a requirement ("requirement: an identifiable element of a function specification that (..)" to that of a specification ("specification: a collection of requirements which (..)").

Moreover, there is a great amount of publications on the art of writing good requirements, revealing the difficulty to produce requirements of good quality. In 1796, Joseph de Maistre had already highlighted in his *Considérations sur la France* that "it is much less difficult to solve a problem, than to define it". This difficulty seems to increase when a natural language is used as a specification means. Natural languages due to their polysemic richness leave the door wide open to all ambiguities, misinterpretations and all that is implied and unspoken. This is what each engineer may experience daily, especially when he/she works in a context spread over several geographic, linguistic and cultural areas.

The question that this observation asks is the following: is the statement of requirements in a natural language (more or less coded), which we call TBR, efficient and for which cost? Basically, are TBRs not too subjective productions to allow a shared understanding between all stakeholders of the problem to be solved? Would it not be better constrain the expression of a requirement until it is completely unambiguous, in other words, until it can be processed by a computer?

We claim that the most efficient way of obtaining requirements, which as IEEE 1220 indicates are unambiguous, testable or measurable, at a reasonable cost, is to give an objective expression of them that does not leave any room for interpretation in the signification of this expression. Moreover, we assert that this objectivity is necessary for an MBSE approach. The MBSE approach of systems engineering poorly accommodates TBR engineering. The reason for this discordance is due to the fact that TBR engineering is consistent with a systems engineering approach that focuses on documents, it shares the same engineering vision, of the same scientific paradigm (using a buzzword that made the fame of Kuhn [KUH 70]).

What we propose here is a formalized approach of requirements, which we called PBRs to emphasize the relationship with the concept of property, widely mentioned in the "foundations" part of this book (see section 1.5).

We consider that this requirement engineering approach is based on the same paradigm as that of the MBSE approach. This is why we define the set of PBRs assigned to an entity such as the specification model of this entity.

Orientation	Requirements	Design
Document oriented	Text-based	Design description
Model oriented	Property-based	Design models

Table 6.1. *Document-centric versus model-centric paradigms*

6.4. Objectifying requirements and assumptions through property-based requirements

We claim that the theory of PBRs is a feasible and economic method for solving problems caused by the subjectivity of text-based specifications and a means for reducing the disorder entailed by this subjectivity in system development processes. The theory of PBRs is a theory of objective specifications.

6.4.1. *Definition*

We call PBR, concerning a type of systems Σ, a constraint enforced on some property P of Σ. This constraint enforces that the values of property P are in a domain D when condition C is fulfilled, provided that Im(P) is the domain of possible values of P and D is a strict subset of Im(P). This definition can be formalized as follows:

$$PBR\text{: [when C]} \rightarrow val(O.P) \in D \subset Im(P)$$

In this expression, the term *PBR* is a label that identifies the requirement; this label is mandatory.

In the optional expression *[when C]* (signaled by the presence of brackets []), the term C denotes the condition of actualization (Boolean expression that may be true or false) that is relevant in the context in which the requirement is expressed. Condition C may denote, for example, a state, an event and a mode (see these terms in section 2.5) of a system belonging to the type of systems Σ, or a state or even an event of its environment.

The term O denotes a concrete object or a type of concrete objects. Object O may be Σ itself or one or several elements of its composition C, environment E or structure S. P is a material property of this object or this type of objects. The domain of possible values of this property is called the image of P and is denoted by Im(P), and thus the domain of values D indicated by the requirement is such that D is strictly included in Im(P) (i.e. $D \subset Im(P)$). Therefore, the expression, which reads as the values of property P of object O (val(O.P)), shall be in domain D, which is a strict subset of the image of P.

Im(P) is a set determined by the definition of property P. If P is a qualitative property, then Im(P) is a finite set $\{a_1, .., a_n\}$; otherwise if P is a quantitative property, then Im(P) may be a countable or an unaccountable part of $\mathbb{R}^n$, $\mathbb{C}^n$ where n is a positive integer.

In summary, the PBR requirement above means that "when condition C is met, the values of property P of object O shall be in the subset D of Im(P)".

Note that the PBR requirement does not say anything about what is to happen when condition C is not satisfied.

Therefore, when condition C is met and the value of O.P is not in domain D (i.e. val(O.P) $\in$ Im(P)-D), we say that the PBR requirement is not satisfied or is breached.

Also note that the requirement:

$$PBR\text{: [when C]} \rightarrow val(O.P) \notin D \subset Im(P)$$

is identical to the requirement:

$$PBR'\text{: [when C]} \rightarrow val(O.P) \in Im(P)\text{-}D$$

In other words, a PBR allows both an obligation and a prohibition to be expressed in the same way.

Moreover, we say that a PBR is an assumption regarding the type of systems Σ denoted by (C,E and S), if O is an object belonging to environment E of this type of systems Σ. This distinction, which was introduced in the initial versions of KAOS [LET 02], and unfortunately, subsequently abandoned, helps differentiate between constraints assigned to the composition and the structure of the system on which designers may act and those assigned to the environment to which the designer is subjected.

6.4.2. *Examples*

Note that $SAT_K(\{PBR_i\}_{i \in I})$ refers to a collection of type K systems that satisfy the set of PBR_i when i belongs to I. For example, we can consider the collection of transport HC helicopters with at least 10 passenger seats, autonomy of at least 200 miles and a direct operating cost less than $1,500, which allows us to define three PBRs:

PBR_1: val(HC.Passenger_seat_number) $\geq$ 10;

PBR_2: val(HC.range) ≥ 200 nm;

PBR_3: val(HC.Direct_Operating_Cost) ≤ $1,500.

We are then able to complete this first specification without any particular problem by adding some PBRs. The only questions that remain to be answered are, on the one hand, whether they are necessary (and from which point of view?) and, on the other hand, will a multiplication of PBRs not cause infeasibility (this concept is already defined in section 6.2).

For example, we can add a maximum rate of climb requirement such as:

PBR_4: val(HC.MaxVz) ≥ 1,750 ft/min

or a maximum velocity such as:

PBR_5: val(HC.MaxSpeed) ≥ 180 mph

Furthermore, when designing a type of systems, we may need to determine important characteristics including, for example, a never exceed speed (VNE) and the takeoff decision point (TDP). Determining these characteristics is imposed by the regulations. However, these requirements are not assigned to the type of systems itself but to the development and certification processes for the type of systems. In other words, these requirements are assigned to the Design Office, which is the designing system[3] for the type of systems considered.

Similarly, other regulatory requirements may constrain the operation of a type of systems, such as the operation rule requiring to never exceed the velocity never exceed (VNE), or the one requiring to reject a takeoff, in the event of one

3 According to the ISO15288 standard, the Design Office is an enabling system for the system of interest (p. 55).

engine failure occurring before the take-off decision point (TDP), and to continue beyond this point to complete the whole takeoff path. These are, therefore, requirements assigned to the operating process for the type of systems considered and are included in pilot's handbooks or flight manuals (here, the enabling system is the pilot).

However, these requirements, assigned to the operation process of the type of systems considered, may in turn be reassigned to systems supporting operations, such as an autopilot, which will never permit the VNE to be exceeded when it is engaged and can be formalized as follows:

PBR_6: when AFCS_engaged → val(HC.AirSpeed) ≤ VNE - threshold

Similarly, continued takeoff when one engine failure occurs beyond the TDP and the emergency procedure is properly applied by the pilot presupposes that the aircraft has the capability to achieve this goal.

PBR_7:when HCtakingOff and OEIafterTDP and OEItakeOffEmergProcedure=OK

→ val(HC.TakeOff_PathPerformance) = Successful

6.4.3. *Typology and sources of PBR*

The previous examples provide us with the opportunity to suggest a typology of PBR that is both justified and coherent and clarify two concepts that are often confused, the type of a requirement and source of a requirement.

In fact, the guides (e.g. SEBOK v1.2, p. 228 [SEB 13]) and standards (ISO/IEC/IEEE 29148:2011, ED79A/ARP4754A) often provide classifications (also called taxonomies and typologies) that mix the two aspects (sources and types) and do not define a typology in the strict signification[4] of the term.

For example, ISO/IEC/IEEE 29148 [ISO 11] introduces the following types of requirements (section 5.2.8.2, p. 14):

– functional;

– performance;

– usability/quality-in-use requirements;

– interface;

– design constraints;

– process requirements;

– non-functional:

  - quality requirements,

  - human factor requirements.

In this context, it is difficult to claim, on the one hand, that an individual requirement shall be complete (section 5.2.5, p. 11) and, on the other hand, characterizing (section 5.2.8.2, p. 14) a performance requirement as "a requirement that defines the extent or how well, and under what conditions, a function or task is to be performed", which means that a performance requirement completes the characterization of a functional requirement (which is, therefore, not complete when it lacks performance attributes).

Similarly, quality requirements, such as the reliability of services provided by a system (for example, the probability of indicating an undetected erroneous altitude will be less than 10^{-9} per flight hour), come clarify and objectify some aspects

4 A typology, strictly speaking, is a mathematical partition of a set E, that is to say any element of E belongs to one and only one partition for each level of the typology.

of a functional requirement, which in their absence, would remain without content (here, indicating altitude). It is, therefore, unusual to place them into a "non-functional" category as is done in ISO/IEC/IEEE 29148 p. 14.

We can provide multiple examples to show that this type of classifications is not robust, and it is not based on well-defined classification criteria, but rather on empirical inventories[5], which in practice lead to damaging situations, where two designers classify the same requirement into different categories, such as "the following flight data (..) shall be registered for maintenance purposes" which may be classified by a designer as a maintainability requirement (and, therefore, non-functional), whereas another may classify it as a functional requirement of a flight data recording system.

Our theory of PBRs clarifies this situation.

We call the sources of a given requirement the stakeholders who express this requirement *vis-à-vis* of one type of systems. These sources may be the acquirer (the client – operator – or its substitutes: marketing and client support of the company) or other stakeholders (certification authority and speciality engineers – certification, safety, costs, environmental conditions, design, production, tests, maintenance and training entities, etc.). It is always useful to be able to relate a requirement to its source without forgetting that the requirement may have multiple sources and that this connection does not predetermine any type of requirement.

In addition, there are three and only three types of PBRs assigned to a system, namely:

– structural;

– behavioral;

5 These are a ransom paid to pragmatism and consensualism.

– mixed (structural/behavioral).

6.4.3.1. *Structural requirements*

We call structural PBR, any PBR that concerns a structural property of a system (see Chapter 1), namely its composition and structure.

The examples of structural PBR include (1) a PBR concerning the presence or absence of such and such a component; (2) a PBR about the shape, composition of a component and its dimensions; (3) PBR on the presence or absence of redundancies, on the number of inputs or outputs and (4) a PBR concerning the physical quantities that characterize its state (mass, inertia, impedance, etc.).

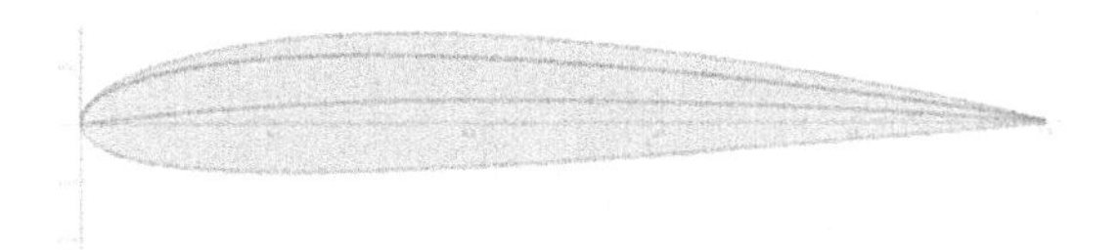

Figure 6.3. *Airfoil structural properties and related PBR*

Therefore, the requirements concerning the following wing characteristics composed of structural requirements, area, thickness, camber, lift and drag coefficients, are ultimately summarized in a National Advisory Committee for Aeronautics (NACA) profile.

6.4.3.2. *Behavioral requirements*

We call behavioral PBR, any PBR about a behavioral property of a system, namely its behavior (most frequently elided in its observable behavior). Thus, PBRs on lift force, drag force or wing speed in an air mass are examples of behavioral PBRs.

Due to the laws of fluid mechanics, behavioral and structural properties of a wing are linked to each other by relations such as $L = 1/2\ \rho C_L S v^2$ where L and v are behavioral properties, whereas ρ, C_L and S are structural properties of the air mass or wing (as defined in section 1.5.3; this interdependence is characteristic of the essential properties of an object). These real links between the essential properties are a condition of possibility of engineering activities. Consequently, couplings between structural and behavioral requirements about the essential properties of an object exist. These couplings between requirements may cause unfeasibility. When this is the case, we say that the corresponding requirements are conflicting.

6.4.3.3. *Mixed requirements*

As we will see in the next section, it is possible to combine PBRs using a conjunction operator, in other words, to define the conjunction of two or several PBRs as a PBR. This brings us to introduce a third type of PBR namely mixed PBR. A mixed PBR is a conjunction of structural and behavioral PBRs.

An example of a mixed PBR would be an interface PBR that fixes both structural (geometric, mechanic and electrical characteristics) and behavioral characteristics communication and transmission speed protocols and error handling) of an interface.

6.5. Conjunction and comparison of property-based requirements

In this section, we are going to introduce two surprisingly new concepts of requirement engineering that are also rather "natural". The first concept is a comparison relationship between PBRs, called "is less constraining than", and the second concept is an operator between PBRs: the conjunction of two PBRs.

6.5.1. *Comparison of two PBRs*

Let us consider the following PBRs: PBR_1: val(HC.Passenger_seat_number) $\geq$ 10, namely a type HC of helicopters, to be developed, shall provide at least 10 passenger places. We can see that this requirement is clearly less constraining than the following requirement: PBR_2: val(HC.Passenger_seat_number) $\geq$ 12, i.e. "that can the most, can the least" and that an aircraft capable of transporting 12 people can, *a fortiori,* only transport 10 passengers.

To express that PBR_1 is less constraining than PBR_2', we can write: $PBR_1 \leq PBR_2$. Similarly, we can say that, if an object of type HC satisfies the requirement PBR_2 then, *a fortiori*, this object satisfies the requirement PBR_1. We can also say that $SAT_K(PBR_2) \subset SAT_K(PBR_1)$. In other words, the less constraining a requirement is, the more objects there are that are capable of responding to it, i.e. logically, the intension of a predicate is the inverse of its extension.

This brings us to the following definition:

$$PBR_1 \leq PBR_2 \Leftrightarrow SAT_K(PBR_2) \subset SAT_K(PBR_1)$$

What results is that the more constraining a requirement is, the more the space of solutions satisfying this requirement is reduced. This makes sense, but deserves to be formalized.

In particular, if $SAT_K(PBR)=\emptyset$, then there is no object of type K that meets this PBR. In this case, we are confronted by an unfeasibility (see section 6.2.).

Similarly, we can say that requirements $\{PBR_i\}_{1\leq i\leq n}$ applied to a type K of systems are incompatible if $SAT_K(\{PBR_i\}_{1\leq i\leq n})= \emptyset$. For example, an electrical harness having to weigh less than P and having to supply electrical power W in 28 VDC to n consumers situated at a distance d from the electrical generator can be impossible to implement, due to the fact that the PBRs on weight, power supplied, power transport distance, voltage to be maintained and the number of consumers to supply can be incompatible.

6.5.2. *Conjunction of two PBRs*

Let us now consider the two following PBRs: PBR_1: val(HC.Passenger_seat_number) $\geq$ 10 and PBR_2: val(HC.range) $\geq$ 200 nm, i.e. that of a type HC of helicopters, to be developed, will have an autonomy of at least 200 nautical miles.

If an object of type HC simultaneously satisfies requirements PBR_1 and PBR_2, we can say that this object simultaneously belongs to the $SAT_K(PBR_1)$ and $SAT_K(PBR_2)$, i.e. HC belongs to the intersection $SAT_K(PBR_1) \cap SAT_K(PBR_2)$. The reciprocal is immediately true. This brings us on to define the conjunction operation between two PBRs as follows:

$$PBR = PBR_1 \wedge PBR_2 \Leftrightarrow SAT_K(PBR) = SAT_K(PBR_1) \cap SAT_K(PBR_2)$$

This definition easily entails that $PBR_1 \wedge PBR_2 \leq PBR_1$ and that $PBR_1 \wedge PBR_2 \leq PBR_2$, that is to say that the conjunction of two PBRs is more constraining than either term since $SAT_K(PBR_1) \subset SAT_K(PBR_1) \cap SAT_K(PBR_2)$ and $SAT_K(PBR_2) \subset SAT_K(PBR_1) \cap SAT_K(PBR_2)$.

We can also establish that a PBR which is more constraining than PBR_1 and PBR_2, is ultimately more constraining than $PBR_1 \wedge PBR_2$. In fact, if $PBR_1 \leq PBR$ and $PBR_2 \leq PBR$, then $SAT_K(PBR) \subset SAT_K(PBR_1)$ and

$SAT_K(PBR) \subset SAT_K(PBR_2)$, and therefore $SAT_K(PBR) \subset SAT_K(PBR_1) \cap SAT_K(PBR_2)$, i.e. $PBR_1 \wedge PBR_2. \leq PBR$.

In conclusion, if $\{PBR_i\}_{i \in [1,n]}$ is the set of PBRs applicable to a type K of technological systems, then the set $\{PBR_i\}_{i \in [1,n]}$ equipped with a conjunction operation $\wedge$ generates a formal structure of a semi-lattice [MIC 08] whose maximum element is provided by $PBR_1 \wedge .. \wedge PBR_n$.

This maximum element constitutes a specification of a subtype of K type systems defined by $SAT_K(\{PBR_i\}_{i \in [1,n]}) = SAT_K(PBR_1 \wedge .. \wedge PBR_n)$.

For example, the set of transport helicopters with at least 10 passenger seats, autonomy of at least 200 nautical miles and a direct operating cost less than \$1500 corresponds to the set: $SAT_K(PBR_1 \wedge PBR_2 \wedge PBR_3)$ with

PBR_1: val(HC.Passenger_seat_number) ≥ 10

PBR_2: val(HC.range) ≥ 200 nm

PBR_3: val(HC.Direct_Operating_Cost) $\leq 1500\$$

Now, if we consider, on the one hand, a PBR assigned to a type K of systems and, on the other hand, a set $S=\{PBR_i\}_{i \in [1,n]}$ of PBRs also assigned to K, it comes below a theorem with a practical importance when dealing with derived requirements and the associated safety problems.

THEOREM.–: *If $SAT_K(PBR_1 \wedge .. \wedge PBR_n)$ is the set of type K of systems that satisfy requirements $S=\{PBR_i\}_{i \in [1,n]}$, then $SAT_K(PBR \wedge PBR_1 \wedge .. \wedge PBR_n)$ is included in $SAT_K(PBR_1 \wedge .. \wedge PBR_n)$. This statement is equivalent to say that $PBR \wedge PBR_1 \wedge .. \wedge PBR_n$ is more constraining than $PBR_1 \wedge .. \wedge PBR_n$.*

This theorem immediately shows that when we add a PBR to a specification S, we tend to reduce the number of possible solutions $SAT_K(PBR \wedge S)$ of this specification.

However, this reduction may not be effective and may give $SAT_K(PBR \wedge S) = SAT_K(S)$. In this particular case, introducing additional PBRs does not contribute anything in technical terms, which, technically speaking, is a noise. Managing this noise may prove to be extremely costly.

Another particular case is given by: $SAT_K(PBR \wedge S) = \varnothing$, i.e. the addition of a PBR to the specification S leads to unfeasibility. The PBR requirement and the specification S are incompatible.

6.6. Interpreting text-based requirements

We call "interpretation" the operation by which TBRs provided by different sources (clients, marketing, administration, client support, manufacture, flight test, etc.) are transformed into one or several PBRs.

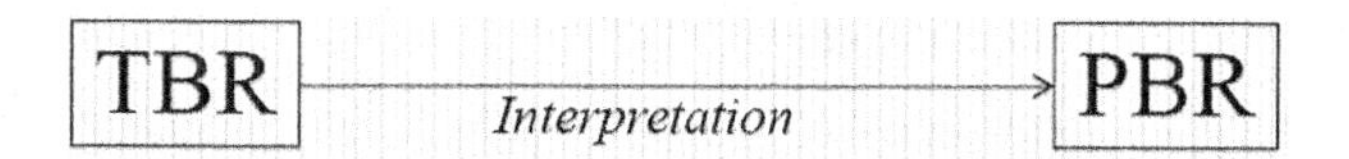

Figure 6.4. *Interpretation: a process from TBR to PBRs*

We assume that this interpretation is always possible in practice, even if it can be very complicated. In other words, it is always possible to clarify unclear statements, although this conjecture cannot be demonstrated. It requires to refer to knowledge items shared (intersubjective knowledge) by stakeholders of this interpretation process, and preferably objective (verifiable) and true knowledge.

In the aeronautical domain, there is a very important interpretative material, including the advisory circulars (ACs), the acceptable means of compliance (AMC), the technical standard orders (TSOs) and European Technical Standard Orders (E-TSOs), the minimum operational performance standards (MOPSs), published, depending on the case, by certification authorities (federal aviation administration (FAA), European aviation safety agency (EASA), transport Canada civil aviation (TCCA), etc.) or aeronautical standardization organizations (EUROCAE, SAE, RTCA, ARINC, etc.). Moreover, when this interpretative material is insufficient, it is still possible for the authorities and manufacturers to agree on a shared interpretation through an Issue Paper for FAA (IP) or a Certification Review Item for EASA (CRI). All elements of this interpretative material are further proof that interpreting TBRs, originally quite vague, may lead to well-formed requirements, even if it is a fairly long process that can still be perfected and revised.

6.6.1. *Example 1: FAR29.1303(b) flight and navigation instruments*

If we consider, for example, the following airworthiness requirement: *FAR 29.1303 Flight and Navigation Instruments. The following are required flight and navigational instruments:*

a) an airspeed indicator. For Category A rotorcraft with VNE less than a speed at which unmistakable pilot cues provide overspeed warning, a maximum allowable airspeed indicator must be provided. If maximum allowable airspeed varies with weight, altitude, temperature, or rpm, the indicator must show that variation;

b) a sensitive altimeter;

c) a magnetic direction indicator.

It is obvious that the term "sensitive" used in this requirement FAR29.1303(b) is particularly subjective and vague. The authorities and manufacturers shall, therefore, agree on an interpretation of what is meant by *sensitive* to create a definition shared by all stakeholders. Nowadays, this definition is found[6] in a MOPS, namely the standard SAE-AS8002A that specifies the minimum operational performances of an air data computer (ADC). It has not always been like this, and we cannot be sure that this definition will not evolve in the future, if needed.

In this standard, for the altimetry of an ADC to be considered compliant with the airworthiness requirement *FAR 29.1303(b),* its accuracy will be within the tolerances provided in Table 6.2:

Altitude (feet)	Altitude (meters)	Tolerance (± feet)	Tolerance (± meters)
0.	0.	25.	8.
1000.	305.	25.	8.
2000.	610.	25.	8.
3000.	914.	25.	8.
4000.	1219.	25.	8.
5000.	1524.	25.	8.
8000.	2438.	30.	9.
11000.	3353.	35.	11.
14000.	4267.	40.	12.
17000.	5182.	45.	14.
20000.	6096.	50.	15.
30000.	9144.	75.	23.
40000.	12192.	100.	30.
50000.	15240.	125.	38.

Table 3.2. *Tolerances on computed altitude by an ADC required by the SAE-AS8002A*

In Table 6.2, we see that the absolute error for the indicated altitude (Indicated_Alt) provided by an ADC installed in an aircraft (AC) should be less than 25 ft when this aircraft is actually situated between 0 and 5,000 ft, for this indicator to be considered to comply with the

6 In the past, other standards have prevailed (SAE-AS 362).

airworthiness requirement FAR 29.1303(b). In other words, expressed in terms of PBR, we will have:

PBR$_1$: when *AC.Alt* ∈*[0.0, 5000.0[* →|*ADC.Indicated_Alt - AC.Alt* | ≤ *25.0 ft*

Moreover, we see that this PBR alone is not sufficient to cover the whole flight domain (beyond 5,000 ft). Thus, we will introduce the following additional PBRs to be able to cover the whole domain considered by the standard:

PBR$_2$: when *AC.Alt* ∈*[5,000.0, 8,000.0 ft[* → |*ADC.Indicated_Alt - AC.Alt* | ≤ *30.0 ft*

PBR$_3$: when *AC.Alt* ∈*[8,000.0, 11,000.0 ft[* → |*ADC.Indicated_Alt - AC.Alt* | ≤ *35.0 ft*

PBR$_4$: when *AC.Alt* ∈*[11,000.0, 14,000.0[* → |*ADC.Indicated_Alt - AC.Alt* | ≤ *40.0 ft*

PBR$_5$: when *AC.Alt* ∈*[14,000.0, 17,000.0[*→ |*ADC.Indicated_Alt - AC.Alt* | ≤ *45.0 ft*

PBR$_6$: when *AC.Alt* ∈*[17,000.0, 20,000.0[*→ |*ADC.Indicated_Alt - AC.Alt* | ≤ *50.0 ft*

PBR$_7$: when *AC.Alt* ∈*[20,000.0, 30,000.0[*→ |*ADC.Indicated_Alt - AC.Alt* | ≤ *75.0 ft*

PBR$_8$: when *AC.Alt* ∈*[30,000.0, 40,000.0[*→ |*ADC.Indicated_Alt - AC.Alt* | ≤ *100.0 ft*

PBR$_9$: when *AC.Alt* ∈*[40,000.0, 40,000.0[*→ |*ADC.Indicated_Alt - AC.Alt* | ≤ *125.0 ft*

We can, therefore, interpret the airworthiness requirement FAR 29.1303(b) as the conjunction of nine PBRs above ($PBR_{FAR\ 29.1303(b)}$ = PBR1∧..∧PBR_9); it is wise to have the following assumption, which assumes that the AC flight domain is situated between 0 and 50,000 ft (above the mean sea level (MSL)), which can be written:

$ASSUMP_0$: $AC.Alt \in [0.0, 50000.0[$

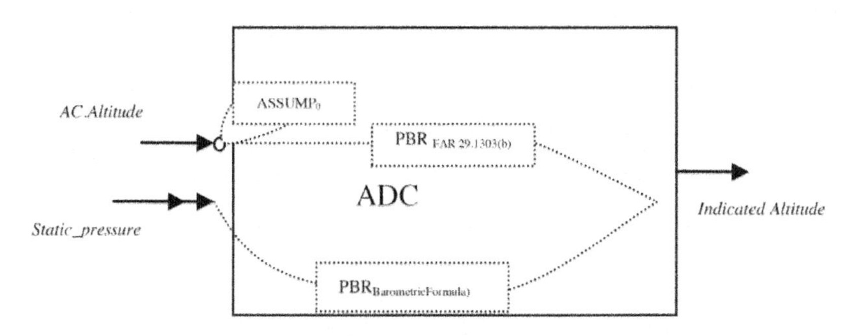

Figure 6.5. *Some PBRs linking inputs and outputs of an air data computer*

Here, we have shown that a requirement can be interpreted as a conjunction of PBRs.

Similarly, we can also specify how indicated altitude has to be computed from static pressure by using a body of knowledge on atmospheric data such as the *Manual of the ICAO Standard Atmosphere* [ICA 64] to express the following $PBR_{BarometricFormula}$:

$PBR_{BarometricFormula}$: when $AC.Alt \in [0\ .11km[\rightarrow$

$|ADC.Indicated\_Alt - BarometricFormula(AC.Pressure)| \leq 25.0$ *with*

$BarometricFormula(AC.Pressure)$

$$= 44330.76923 * \left[1 - \left(\mathrm{AC.Pressure}/760\right)^{0.190263}\right]$$

6.6.2. *Example 2: FAR29.951(a) Fuel systems – General*

We can also consider another airworthiness requirement that is also subject to interpretation. This is the general requirement FAR29.951(a). We have chosen this as it takes place in the domain of continuous systems (fluids).

FAR 29.951 Fuel systems – General

a) Each fuel system shall be constructed and arranged to ensure a flow of fuel at a rate and pressure established for proper engine and auxiliary power unit functioning under any likely operating conditions, including the maneuvers for which certification is requested and during which the engine or auxiliary power unit is permitted to be in operation.

This requirement includes helicopter fuel systems subject to FAR 29. It refers to an essential property of an aircraft fuel system, namely to supply fuel to the associated engine. The statement *"the fuel system shall supply fuel to the engine"* it not itself a requirement, since a system that does not supply fuel cannot be called a fuel system (here, the functional requirement concept is useless). The requirements are the modalities according to which the system shall supply the fuel, as requirement FAR 29.951 indicates, to provide fuel (continuously) to the assigned engine, at the flow rate and pressure required by the engine, under all operation conditions (including authorized operations), and throughout the whole duration of the authorized operation.

$PBR_{FuelFeed}$: when *Ctrl= Feeding* $\wedge$ *c in Operation_conditions* $\wedge$ *t in Operation_time* $\rightarrow$ |*Fuel_System.Fuel_Rate&Pressure (c,t)– Engine_Demand(t)*| $\leq$ *tol*

This requirement is read as follows: when the system is in feeding mode, and the operation conditions (altitude, temperature, etc.) remain in a defined domain, and the operation time does not exceed a defined time, then the fuel system shall supply the continuous flow of fuel required by the engine with an acceptable pressure level.

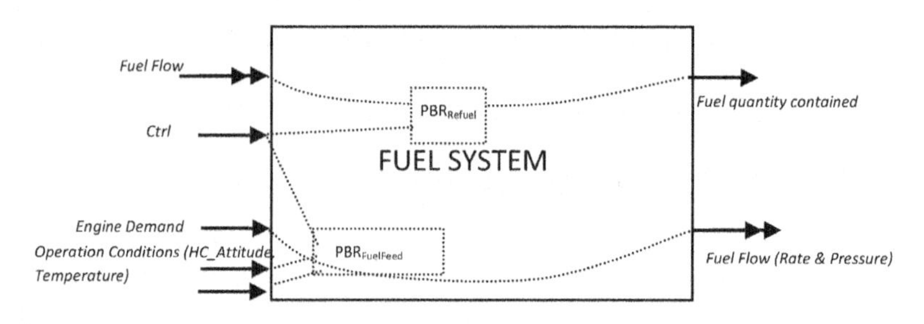

Figure 6.6. *Some PBRs linking inputs and outputs of an aircraft fuel systemE*

We can also introduce a requirement for the filling of the system with fuel, which links the quantity of fuel contained in the system and the quantity of fuel supplied when the system is in *refueling* mode.

PBR_{Refuel}: *when Control= Refueling* $\rightarrow$ $| \, | Fuel\_System.Fuel\_Quantity\ (t_f)\ Fuel\_System.Fuel\_Quantity\ (t_i)\ | - \int_{t_i}^{t_f} Fuel\_Flow(t)dt\ | \leq tol$

In other words, when the system is in refueling mode, the difference between the contained quantity at the end of filling and that contained at the start shall be equal (within a stated tolerance, to account for acceptable losses) to the quantity supplied to the system.

Additional requirements can be added to these two requirements, for example, a maximum refueling time and others about the fuel gauging accuracy, reliability, etc.

To conclude on the interpretation of TBRs, note that many TBRs keep implicit the object to which they relate. Also, many system specifications contain requirements that do not relate to itself, but to the associated processes such as the development process (for example, "software assurance level of the SAS[7] shall be A in accordance with DO-178C" means that the software development process of the SAS software will satisfy the objectives required by DO-178C for the level A) or the system operating process (such as weight and balance requirements that the operator will respect when cargo is loaded on board an aircraft).

6.7. Conclusion: specification models and concurrent assertions

A system is usually represented graphically in the form of a rectangle with inputs and outputs as shown in Figure 6.7.

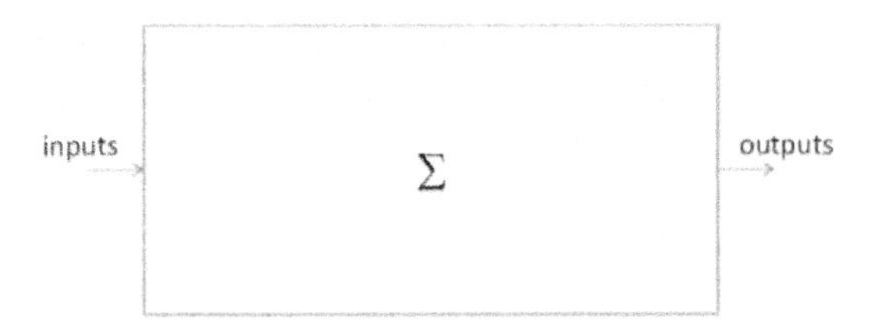

Figure 6.7. *Canonical graphical representation of a system*

This summary representation of a system appeals to our imagination but it disregards several facts:

First, this representation does not always, unlike a schematic representation, provide a true physical image of the system's physical bonds. Thus, the feed line (physical link) that links the fuel system to an engine (in a schematic representation) is represented here as an input (for example, an requested flow rate which is a first charachetristic of the fuel line.) and an output (for example, a fuel pressure exerted which is the second charachetristic of the same fuel line- input

7 SAS: stabilization augmentation system.

and output provide the two variables of a power bond.). In fact, what is shown in this canonical representation is a sequence of causes and effects. Represented to the left of the rectangle are the causes (inputs), whereas to the right are the effects (outputs), visible effects of a system, consequences of actions suffered by the system and its own activity [THO 75].

Second, most systems have a memory (like human beings) and the outputs at a given moment in time not only depend on the inputs at that moment but also on all previous inputs (in the case of causal systems).

To account for this, there are two possibilities:

1) Either to take into account the inputs from initial conditions up to the current moment to determine the outputs at that moment. This is what specialists of dynamic systems or automatic control do (or did), when they determine the transfer function between inputs and outputs. In this case, this takes place in the frequency domain rather than in the temporal domain.

2) Either to adopt what the specialists of dynamic systems or automatic control [PAL 14, KAR 06] call representation by states, in which the intended states from a given input at a given instant are determined by a state equation, whereas the outputs are determined by an output equation. This takes place in the temporal domain.

For the specification model of a type of systems, the second point of view has been adopted rather than the first; the temporal approach is much more intuitive and natural than the frequency approach to most engineers. This causes the specification model of a system (that makes the inputs, observable states and outputs) to be not in biunivocal correspondence with the inputs and outputs of physical systems as shown below. If the inputs and outputs of the specification model correspond to the inputs and outputs of specific physical systems by regrouping inputs and outputs[8], the observable states of the specification model correspond to internal states that have been "externalized" for modeling purposes and to make the model observable.

In this context, the specification model of one type of systems would be equivalent to the conjunction of two PBRs, the first specifying the intended state of the type of systems, as in the state equation, and the second specifying the intended outputs of one type system, as in the output equation.

The specification model of a type Σ of systems can be represented graphically as follows:

1) A rectangular box whose inside is not visible (black box) represents the specified type of systems. In this approach, no information about the composition or endo-structure of type Σ of systems is provided.

2) To the right of the box, outputs and observable states may appear. Among the outputs, the intended effects (or functions) can be distinguished from the concomitant effects, or secondary outputs, as Hubka called them in his theory of technical systems [HUB 84]. These secondary outputs can range from vibrations produced, shocks, electromagnetic emissions (EMC/EMI), etc. Depending on the choice above, there are observable states of the type of systems considered (indications, alerts, statuses, etc.).

3) To the left of the box are continuous or discrete inputs, which represent continuous flows (matter and energy), or discrete flows (data and signals). We can also differentiate the inputs needed for the system to operate and, on the other

8 Thus, a shaft bond involves both an input (for example, rotational speed of the shaft) and an output (transmitted torque), whereas an electrical signal can be a twisted pair of electric cables.

hand, disruptive inputs such as high-intensity radiated fields (HIRFs), lightening, extreme temperatures, etc.

4) Assertions (Boolean functions) linking any number of outputs, observable states and inputs. These assertions represent PBRs. These assertions may link one or several inputs together. In this case, we call assumptions such PBRs (to use a terminology formed by the initial versions of KAOS [LET 02]).

5) As mentioned above, a PBR is an expression in the form PBRn: [when C] $\rightarrow$ val(O.P) $\in$ D. Such an expression is logically equivalent to the assertion "val(O.P) $\in$ D or not (C)" or even to O.P(C,D) where O.P(C,D) is a Boolean function.

6) In simulation, assertions are evaluated according to the context. An assumption about an input should be assessed every time this input changes, whereas a requirement linking one input and one output should rather be assessed every time the output is active (to account for the flow propagation time through the system). Generally speaking, it is up to the specifier to decide in which events, input or output, the requirements will be assessed.

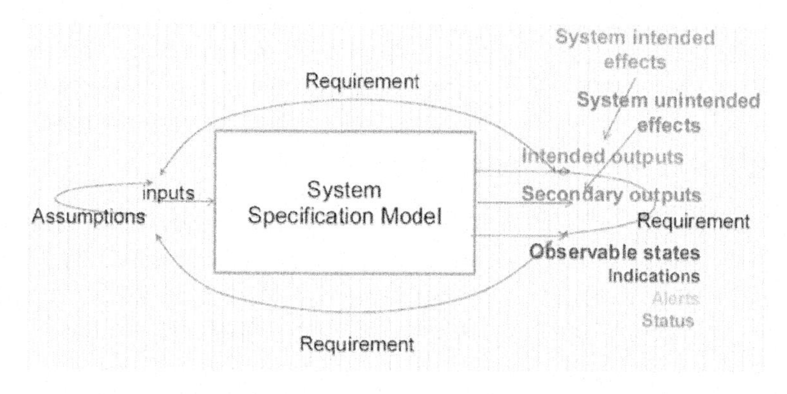

Figure 6.8. *System specification model. For a color version of this figure, see www.iste.co.uk/micouin/MBSE.zip*

When a statement O.P(C,D) (for [when C] $\rightarrow$ val(O.P) $\in$ D), condition C is first assessed:

– if condition C is true, then the function returns the result of expression val (O.P) $\in$ D, i.e. the predicate val (O.P) $\in$ D is assessed and its value (true or false) is returned;

– if condition C is false, then the function returns the constant true value; in other words, the predicate is not assessed.

During the simulation process, the assertions that make up a specification model of a type Σ of systems are assessed in parallel. This means that, at each simulation cycle, all assumptions (requirements or assumptions) that shall be assessed, because the associated triggering events have occurred, are effectively evaluated and the concurrent assessment results are plotted by the simulator. In one cycle, one or several requirements or assumptions may be breached, indicating either an error in the specification or an error in the design, as we will study later on in Chapters 8 and 9.

7

Designing Solutions and Design Models

7.1. Introduction

As mentioned in the introductory Chapter 5, the composition of a technological system is made up of building blocks that its endo-structure assembles (see Figure 5.2, Chapter 5). The same design process is applied for each building block, except for terminal blocks that (1) do not require design, (2) are designed somewhere or (3) have already been designed.

The design process of each block is composed of two subprocesses:

1) a subprocess defining the requirements assigned to the block under consideration;

2) a subprocess defining a solution.

In Chapter 6, we described the process of definition of the requirements assigned to the type of systems targeted. In this chapter, we will discuss on the second subprocess, that is to say defining one or more solutions.

We know that design problems are "inverse" problems, meaning that they can have multiple solutions, which is why we use the expression "definition of one or more solutions" rather than "definition of the solution".

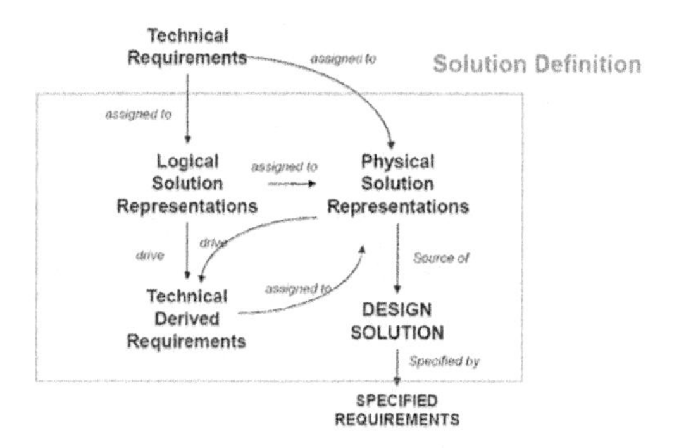

Figure 7.1. *EIA632 building block solution definition subprocess*

The definition of a design solution involves multiple activities: (1) design of a logical solution, (2) allocation and derivation of part of the technical requirements of the system to the logical solution under consideration, (3) design of a physical solution, (4) allocation of the logical solution, and of the requirements allocated to the logical solution, to the physical solution under consideration and (5) allocation and derivation of the pending technical requirements of the system to the physical solution under consideration. As indicated in Figure 7.1, the design subprocess leads to a design solution determined by the specified requirements of the subsystems.

However, before describing the subprocess of defining a solution, we will introduce the concept of derivation of property-based requirements (PBRs).

7.2. Deriving requirements

We propose a definition of the concept of derivation of requirements in this section. The definitions of this concept in standards show little homogeneity.

For the EIA632, a derived requirement is unambiguously a requirement "that is further refined from a primary source requirement or a higher-level derived requirement"[1]. For example, a subsystem requirement can be derived from a system requirement.

For ISO15288, the definition of derived requirement is missing.

For ARP4754A, derived requirements are "additional requirements resulting from design or implementation decisions during the development process which are not directly traceable to higher-level requirements"[2]. However, problems are caused by the ambiguity of natural languages, what does "not directly traceable to" mean? It could mean, for example, that a derived requirement is an additional requirement resulting from a design choice (DC) and is traceable to a higher level requirement through the mediation of (therefore, indirectly) this DC. This is the interpretation provided by Scott Jackson[3], John McDermid [MCD 03] and apparently the ARP 4761 [SAE 96]. It could mean, on the contrary, that a derived requirement cannot be linked to a higher level requirement. This is the chosen interpretation by many practitioners who sometimes campaign for the suppression of the adverb "directly" of the recommendation. We do not share this second point of view that does not rely on an explicit requirement theory and we adopt the point of view of EIA 632, Scott Jackson, etc., which

1 EIA632, page 64.

2 ED791/ARP4754A, page 7.

3 Scott Jackson who speaks, in a very suggestive way, of "derived requirement harmonics" in systems engineering for commercial aircraft, Ashgate Publishing Company, p. 43, 1997.

is consistent with the signification of the word 'derivation' and our PBR theory.

In the theory of PBR, we decide to define the requirement derivation as a mechanism which, during the design process, enables the substitution of a PBR of system level by a conjunction of requirements (PBRs) assigned to its subsystems.

For such a derivation $Rq \rightarrow \{Rq_1,..,Rq_n\}$ to be true, the following formal relation shall be established:

$$\text{when DC} \rightarrow \text{Rq} \leq \text{Rq}_1 \wedge .. \wedge \text{Rq}_n \qquad [7.1]$$

Relationship [7.1] means that the conjunction of derived requirements Rq_i assigned to the subsystems ($s\Sigma_1$, , $s\Sigma_n$) of a system Σ shall be more constraining than the system requirement Rq from which they derive, provided that the DC has been respected.

For example, consider a system Σ with an electric consumption that shall be less than or equal to 1 kW when it is running (Σon). This immediately gives as a PBR

Rq: when Σon → Σ.Electrical_Consumption ≤ 1 kW

Suppose that the two alternate designs DC_1 and DC_2 are considered for the system Σ and the two designs are based on the two same components IT_1 and IT_2. The only difference is that in the first case (DC_1 architecture), IT_1 and IT_2 work in alternation (passive redundancy, for example), whereas in the second case (DC_2 architecture), IT_1 and IT_2 work simultaneously (active redundancy).

Finally, consider the two following requirements:

Rq_1: when IT_1.on → IT_1.Electrical_Consumption ≤ 5.0 kW

Rq_2: when $IT_2.on \rightarrow IT_2.Electrical\_Consumption \leq 5.5$ kW

and the two following derivations:

1) when $DC_1 \rightarrow Rq \leq Rq_1 \wedge Rq_2$

2) when $DC_2 \rightarrow Rq \leq Rq_1 \wedge Rq_2$

In case 1, the derivation $Rq \rightarrow \{Rq_1, Rq_2\}$ is true while in case 2 it is false.

This simple example shows that:

– first, to be able to declare validated the derived requirements Rq_1 and Rq_2, it is necessary to take into account the DC (DC_1 and not DC_2);

– second, to be able to declare validated Rq_1 (respectively, Rq_2), Rq_2 shall be considered (respectively, Rq_1).

In other words, derived requirements cannot be validated one at a time (individually) but only collectively with respect to a DC.

7.3. Basic system model of a type of systems

The property-model methodology approach is based on different types of models. In this way, we have already introduced specification models (SSMs).

Now we will introduce two types of design models:

– static design models, which are divided into:

- structural design models (SDMs) described in section 7.6.2;

- reliability design models (RDMs) described in section 10.3.6.

– dynamic design models, which are divided into:

- equation design models (EDMs) described in section 7.4.2;

- behavioral design models (BDMs) described in section 7.4.1.

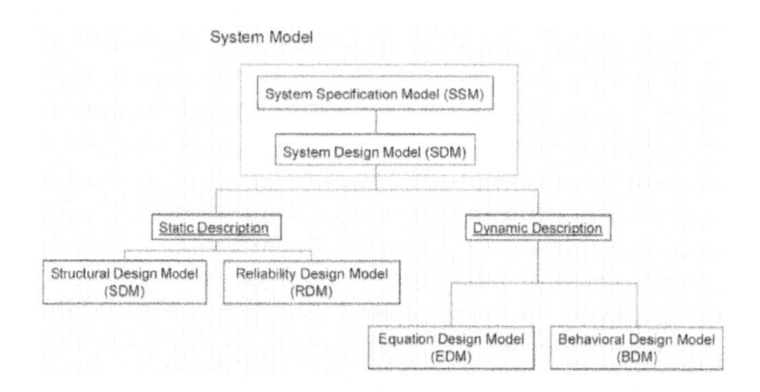

Figure 7.2. *Design model typology*

A basic system model of a type of systems is an assembly built by a specification model and a dynamic design model (see Figure 7.3). The specification model carries the requirements assigned to the type of system, whereas the dynamic model is a theoretical model of the type of systems under consideration, as defined in Chapter 4, section 4.4. It is a model that can be executed ("ran") in the framework of a conceptual simulation (in ones head) or a physical one (on a simulator). When compiled, a system basic model is syntactically true and shall be sufficiently complete to be executed in a simulation environment.

When the system basic model is executed in a simulation, its environment injects inputs (simulation scenarios composed of simulation cases) in the systems model. Outputs are then computed by the design model while the specification model ensures that the inputs and outputs are consistent with the PBRs assigned to the type of systems. If this is not the case, the simulator reports requirement breaches. The simulator thus ensures that (1) the assumptions made about the environment are satisfied (or dissatisfied), (2) the outputs or observable states formulated are consistent (or inconsistent) with the requirements that link the outputs, observable states and inputs, and (3) the requirements focused only on the outputs, or observable states, are satisfied (or dissatisfied).

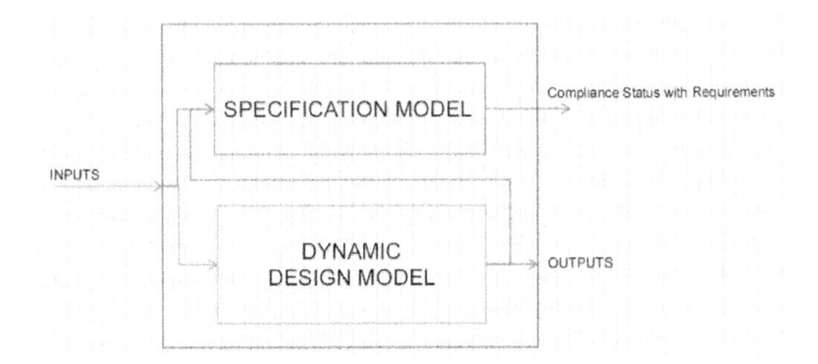

Figure 7.3. *Basic system model. For a color version of this figure, see www.iste.co.uk/micouin/MBSE.zip*

In section 7.6.1, we introduce the notion of composite model systems.

7.4. Dynamic design models of a type of systems

We can distinguish between two types of dynamic models: the BDM and the EDM.

7.4.1. *Behavioral design model (BDM)*

A BDM of a type of systems is a model, which describes the behavior of the targeted type of systems in terms of processes and flows, organized following the architecture chosen by the designer. This constitutes a behavioral architecture[4] of the type of systems targeted.

This behavioral architecture ensures the interconnection between the outputs and inputs of the targeted system so that these inputs and outputs are defined in the corresponding specification model. The BDM is linked to its environment (exo-structure) by the intermediary of the inputs and outputs of the specification model, which are interfaces *vis-à-vis* the exterior world.

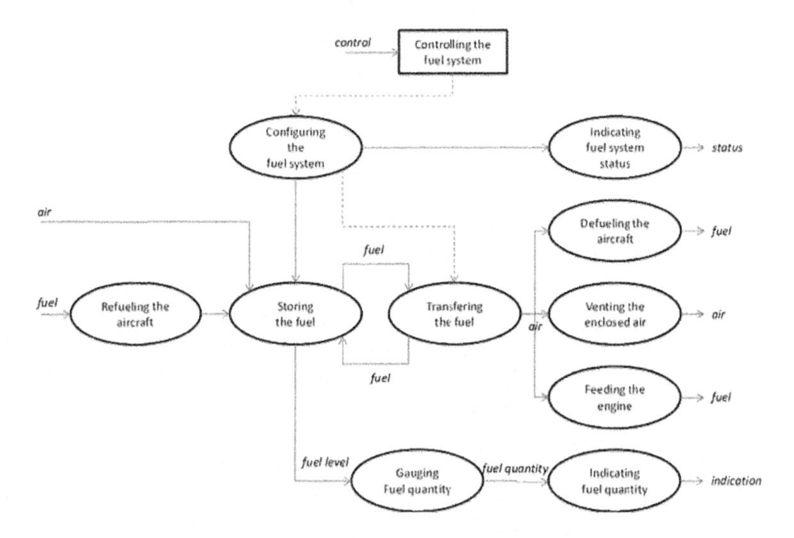

Figure 7.4. *Fuel system behavioral design model*

In simulation context, the processes composing the behavioral architecture are executed concurrently, in a coordinate manner, in accordance with the synchronizations described (possibly in an erroneous manner) by the designer, and by producing and consuming matter, energy, data and signal flows that interconnect them. These processes can be

4 In reference to Gero's FBS framework, we prefer the expression "behavioral architecture" to the generally used expression "functional architecture".

of three types: discrete, continuous or mixed. When discrete, they produce discrete data or signals from discrete data or signals when the events they expect occur. When continuous, they produce waves of continuous matter, energy or data from continuous inputs. When mixed, they produce discrete outputs (respectively, continuous) from continuous inputs (respectively, discrete).

This is, indeed, a synthesis question (and not an analysis one), as the idea is to create or reuse[5] elements (processes and state machines) and a structure (flows), which could allow the model to provide the required outputs as an effect of the inputs (and present state). The objective is, therefore, not to analyze an existing situation but to design[6] a new one.

During the design of a type of systems (characterized by the specification model) we can equip it (within the framework of the search for a preferred solution) with different alternatives of behavioral architecture, each one corresponding to a behavioral design model. Among them, there are those that do not satisfy the requirements of the specification model and do not constitute logical solutions to the specification problem. We will see that the verification process described in Chapter 9 enables us to put them aside. The others, which satisfy the requirements of the specification model, are acceptable logical solutions to the specification problem.

Generally, to design a behavioral architecture, we start by giving a description of the modes (for definition, section 2.5)

5 For example, exterior models available in existing catalogs or produced during preliminary studies (see section 11.1).

6 In accordance with Descartes' third rule of method, "the third was to direct my thoughts in an orderly manner, by starting with the simplest and most easily known objects in order to move up gradually to knowledge of the most complex" Discourse on Method, René Descartes, Part 2, page 9, translated by Jonathan Bennett, 2007, http://www.earlymoderntexts.com/.

of the type of systems under consideration, which means describing the phases of the functioning cycle of the type of systems, and the transitions that enable us to switch from one mode to the other. Described in design as state machines, the modes and transitions between modes have normally been identified in the specification model, either as requirement activation conditions, or as observable states, because these are the modes that temporally divide the targeted systems intended behavior.

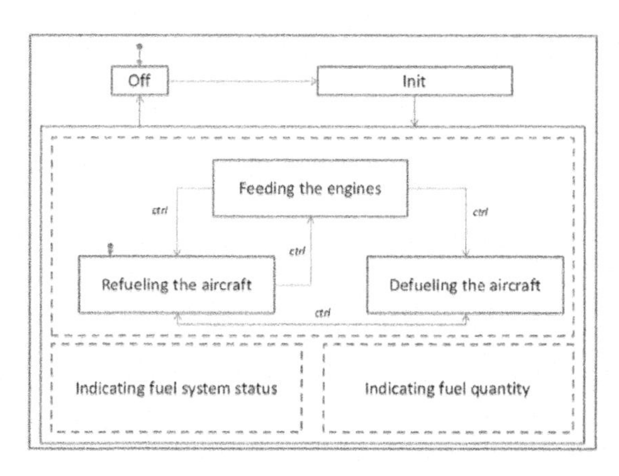

Figure 7.5. *System mode model*

For example, consider a fuel system. It can be at a standstill; in the "off" mode, it produces none of the intended effects. Once started, it passes to an "on" mode after an initializing phase ("init") during which it undertakes the power-on built-in test (PBIT) of different system components (valves, pumps, gauges and calculators) and displays a status to the crew. Once this initializing phase is successfully carried out, it works according to different modes (three), the main one ensures fuel feeding to the engines at the required pressure. The second ensures refueling, while the third ensures the inverse operation (defueling), which involves emptying the tank, during maintenance, for example. In these three exclusive modes (if the filling of tanks during the flight is excluded by specification), the system, on the one hand, carries out continuous built-in tests (CBITs) and presents a status to the crew, and on the other hand, presents the available quantity of fuel on the basis of fuel gauging process.

The state machine representing the functioning modes of the system enables the description of the systems configuration change command (opening/closing of valves, start up/stop of the pumps, etc.) ensuring the activation of certain processes (feeding) and the inhibition of others (refueling and defueling).

The behavioral architecture organizes chains of processes[7] around one or more state machines ensuring the systems main functions in its different modes.

For example, for functional modes of the fuel system mentioned above, we can distinguish:

1) refueling;

2) defueling;

3) feeding the engines;

4) jettisoning the fuel (that does not appear here).

For example, the functional chain "feeding the engines" that enables engine feeding ensures a continuous flow, which goes from the process "storing the fuel" to the process "feeding the engines" passing through the process "transferring the fuel". This chain is activated by the process "controlling the fuel system" that configures the system in order to activate the chains that work in parallel to "feeding the engines" process, that is to say the chains "indicating the fuel system status" and "indicating fuel quantity" and to inhibit the "refueling" and "defueling" chains.

7 These are often called "functional chains".

The "feeding the engines" mode is illustrated below.

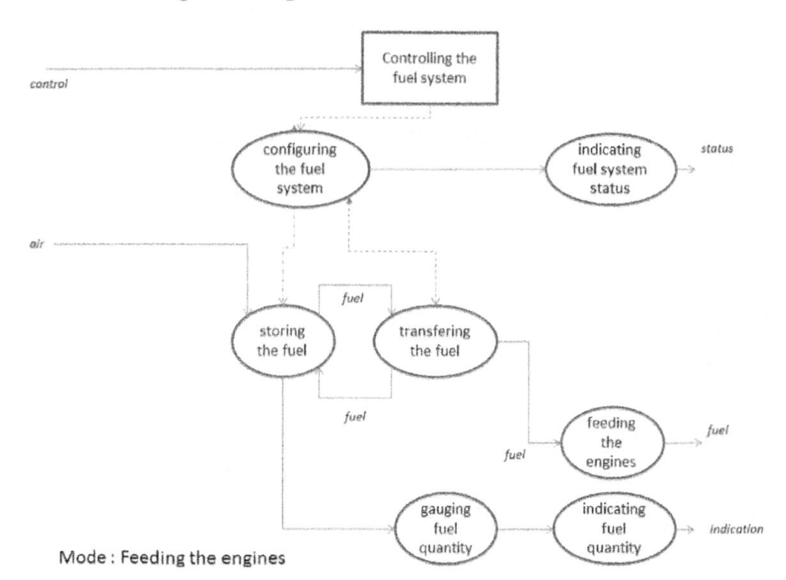

Figure 7.6. *"Feeding the engines" functional chain*

The three functional chains, mentioned above, are synthetically integrated into the behavioral architecture presented in Figure 7.4.

The fuel system is a fluid system with properties, both local (density, pressure, temperature and speed) and global (fuel quantity and gravity center (GC) location) evolving in accordance with the laws of fluid mechanics and thermodynamics (conservation of mass, Newton's second law and conservation of energy). To each element of the model (process), we can associate a specific theoretical model, that is to say a system of equations linking the inputs and outputs of each process, for example, the laminar or turbulent flow equations in a pipe, across a transfer pump or a control valve, as a function of the considered fluid's characteristics, compressible/incompressible and viscous/ non-viscous.

7.4.2. *Equation-based design models (EDMs)*

The prerequisite to carry out a design task is to dispose of a validated specification model of the type of systems to be designed. This means that such a specification model has first been defined and then validated. This notion of validated specification model and the process of validation associated will only be discussed in the following chapter. For the time being, we assume that a specification model is validated if it correctly and completely characterizes the type of systems targeted. In this case, any concrete system conform to the validated specification model is indeed within the set of targeted concrete systems and puts none aside.

For example, for a fuel system, we can start by a basic theoretical model with the aim to validate (this question of specification model validation will be tackled in Chapter 8) the specification model. This minimal design model enables the representation of the intended behavior but, in no way, the effective behavior[8] insofar as it ignores the characteristics of the physical architecture of an aircraft fuel system. This physical architecture is the result of trade-off relative to the number of tanks, the capacity, the shape and position of the aircraft in each of them, which are not free variables but take into account the multiple constraints regarding installation, safety and GC location imperatives.

A precise behavioral model of the fuel system will be established once a physical architecture is defined, in order to verify (this question of verification will be tackled in the Chapter 9) that the design solution correctly answers the requirements assigned to the type of systems.

8 For the distinction between intended behavior and effective behavior, see Chapter 2 (section 2.3).

Among the acceptable logical solutions we have considered in the previous section some are minimal in the sense that they introduce a minimal number of processes to interconnect the inputs and outputs of the specification model, while satisfying the requirement of the model. We call EDM a minimal logical solution. This minimal number of necessary processes to interconnect inputs and outputs of the specification model depends on the number of requirements linking the outputs to the inputs or outputs and also to input flow convergences.

An EDM model is a behavioral model that does not take into account any physical architecture or installation constraints and that confines itself to produce an interconnection of inputs and outputs of the specification model by satisfying the requirements of the corresponding specification model. An EDM model is, therefore, an exact executable image of the specification model from which it is deducted. Such EDM equations can be automatically generated from the specification model.

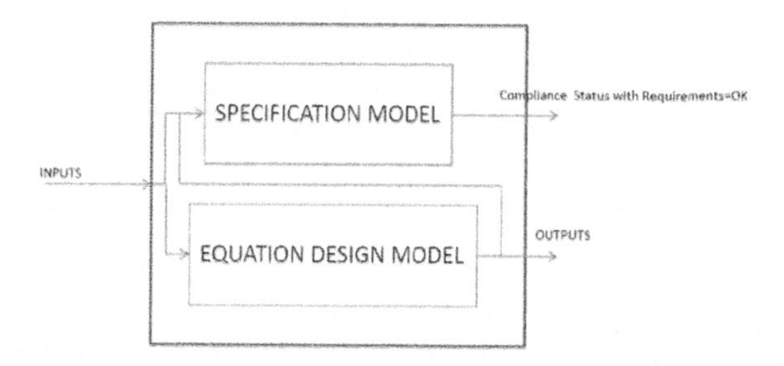

Figure 7.7. *Basic system model including an equation design model. For a color version of this figure, see www.iste.co.uk/micouin/MBSE.zip*

When a basic system model, constituted of a specification model and an EDM model, is executed during a simulation, the result of the execution being conformed to the specification. It is then possible to ensure that it is correct and complete; in other words, the production of an EDM model is a simple way of validating a specification model (see Chapter 8).

7.5. Derivation and allocation of the system's behavioral requirements

Once a behavioral architecture is established for a type of systems, we can then consider the way in which the behavioral requirements of the specification model will project themselves on this architecture by determining the intended performance of each process. Some system requirements could then be put aside (such as mass requirements) and will be taken into account at the structural level. However, for those that are taken into account, this projection consists of deriving the system-level requirements into derived requirements assigned to the different components of the behavioral architecture.

For example, if we want to enhance the availability of an aircraft, the time taken to refuel necessarily becomes a property constrained by a requirement. This system-level requirement "at determinate conditions, the time required to fuel up shall be inferior to x" will then derive itself into performance requirements assigned to the processes of fuel storing, fuel transfer and air flushing replaced by fuel in the fuel system.

To ensure the system-level requirement is satisfied, the derived requirements allocated with the processes of the behavioral architecture must have been correctly characterized, and none must have been forgotten. If this is not the case, that is to say the derivation of the system requirement is incorrect or incomplete, there is a major risk that the system requirement is not satisfied in certain circumstances. This is the reason why a process of validation of the specified requirements (to the behavioral architecture processes), with regard to the requirements of the type of systems, is set up and described in Chapter 8 (see section 8.3.4).

7.6. Static design models

We can distinguish between two types of static design models. We will refer to the first model as SDM (structural design model) and the second as RDM (reliability design model). In this chapter, we only take interest in the SDM model, while the RDM model will be considered in Chapter 10.

7.6.1. *Composite system model*

Previously, we have only considered basic system models, that is to say models constituted of a single system specification model (SSM) and a unique design model (BDM or EDM). These basic models are not enough when we wish to model large and complicated systems.

The SDM is the means by which it becomes possible to represent a system as being constituted of different subsystems that are themselves constituted of items, and so on and so forth, with the needed level of depth. In this way, it is a practical application of the famous "divide and conquer" rule[9], which involves, in the world of technological design, subdividing a whole into a network (endo-structure) of parts to be able to tackle and master it. This does not entail a reduction of the whole to its parts as, of course, the endo-structure ensures the emergence of the properties of the whole relatively to the properties of the parts (see section 1.5.4).

---

9 "This is the second rule of Descartes method". Disourse on Method, René Descartes, Part 2, page 9, translated by Jonathan Bennett, 2007, http://www.earlymoderntexts.com/.

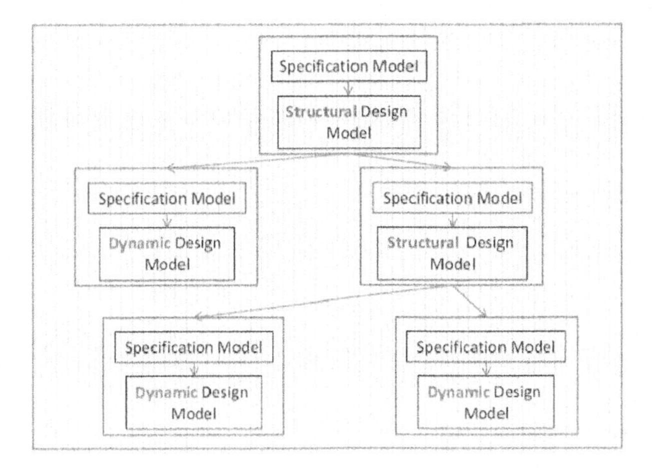

Figure 7.8. *Composite system model. For a color version of this figure, see www.iste.co.uk/micouin/MBSE.zip*

A composite system model is an assembly constituted of specification and design models, and organized according to a tree (tree system). At the root of the tree, we find the node of the type of systems being developed, constituted of its specification model and a structured design model. This SDM refers to subsystem models, which are themselves constituted of a specification and design model. If a subsystem is associated with a non-terminal node in the tree system, then the design model attached to that node is necessarily an SDM. Conversely, if a subsystem model is associated with a terminal node (a leaf) in the tree system, then the design model attached to the node is necessarily a dynamic design model (BDM or EDM).

Following the example of the basic system model, a composite model can also be executed within the framework of a simulation that is conceptual (in ones head) or physical (using a simulator). Compiled, a system composite model is syntactically correct and can be executed in a simulation environment. Particularly, the all the interface compatibility problems, both internal and external (with the environment), are entirely solved during the compilation. Those who know the finalization costs of interface requirement specification (IRS) and interface design description (IDD) will appreciate this.

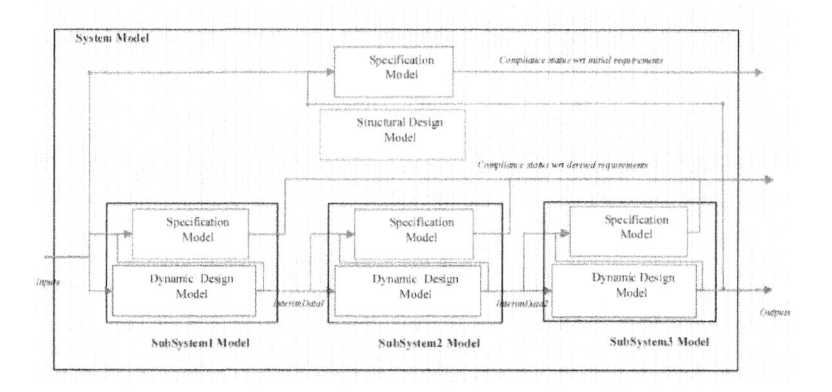

Figure 7.9. *Composite system model. For a color version of this figure, see www.iste.co.uk/micouin/MBSE.zip*

When the system composite model is executed in a simulation, its environment injects inputs (simulation scenarios composed of simulation cases) in the system model. The SDMs connect the inputs and outputs to the subsystem models (not represented in Figure 7.9) up to the terminal nodes. These terminal nodes behave like system basic models, as described in section 7.4. The outputs are formulated by the BDM, while the specification model ensures that the inputs and outputs are in accordance with the PBRs allocated to the type of systems. If this is not the case, the simulator reports the requirements that are breached.

This monitoring process of the consistency of inputs, observable states and outputs with the PBRs is carried out at each level by the corresponding specification models up to the highest level; in other words, each specification level ensures that the inputs and outputs are consistent with the requirements it holds.

We, therefore have, level by level, a complete design model verification process in relation to the corresponding specification models.

7.6.2. *Structural design model*

An SDM of a type of systems describes the structure of the type of systems being targeted. This description is expressed, in terms of subsystems, hardware, items, etc., and, on the other hand, in terms of link variables (piping variables, cables variables, electric harnesses variables, data links and signal links) organized according to a chosen architecture by the designer, and possibly erroneous. In this way, they constitute a physical architecture of the type of systems targeted.

This physical architecture ensures the interconnection between the outputs and inputs of the type of systems targeted such that these inputs and outputs are defined in the corresponding specification model. The SDM is linked to its environment (exo-structure) by the intermediary of inputs and outputs of the systems model; these are its interfaces with regard to the external world.

Thus, the fuel system we described previously in behavioral terms can be described as a network of tanks, vent hoses, feeders, pumps and sensors linked to a computer. Each of the physical components contributes to the creation of processes identified in the behavioral design model. In this way, the tanks ensure the fuel storing. The apertures ensure fuel filling, piping and pumps ensure the transfer of fuel and air flushing events within the tank or on the contrary its intake, the feeders ensure the tube feeding of engines with fuel, and so on and so forth.

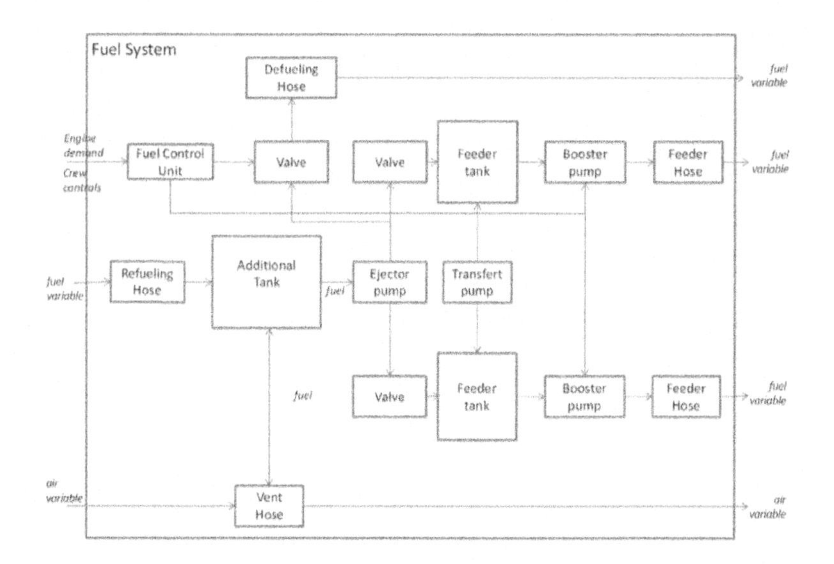

Figure 7.10. *Fuel system structural design model*

7.6.3. *Allocation of BDM components to SDM components*

This allocation of elements of a BDM to the elements of an SDM involves the allocation, on the one hand, of processes to processors and, on the other hand, of flows to link variables. We will then, for example, allocate the process of storing to one or more tanks, the processes of transfer to pipes equipped (or not) with pumps.

Two constraints are imposed for these allocations: first, a process cannot be assigned to multiple processors without a preparatory decomposition and, second, each processor must have at least one process assigned to it.

7.7. Derivation and allocation of system requirements

The allocation of behavioral design elements to physical design elements is doubled by the allocation of behavioral requirements carried by behavioral elements to the physical elements supporting them. We could say that the behavioral requirements are inherited by the processors on a physical level.

Then remains the question of structural requirements of the type of systems, which must also be projected onto the SDM. This projection involves the derivation of structural requirements at the system level into derived requirements allocated to the different components of the physical architecture.

For example, if a fuel system must be capable of boarding a given volume of fuel, and if the design introduces *a given* number of tanks, then this system level structural requirement will be derived into as many structural requirements allocated to the different tanks.

To ensure that the system-level requirement will be satisfied, the derived requirements allocated to the components of the physical architecture must have been correctly characterized and none forgotten. If this is not the case, that is to say if the system requirement derivation is incorrect or incomplete, there is significant risk that the system requirement will not be satisfied under certain conditions. This is the reason why a process validating the specified requirements (to the components of the physical architecture) in relation to the requirements of the type of systems is set up and described in Chapter 8.

As we have explained, this process can be repeated for each building block until encountering either a commercial off-the-shelf (COTS) block or a reusable one, a directly codable block (software) or a directly producible block not requiring any other design task.

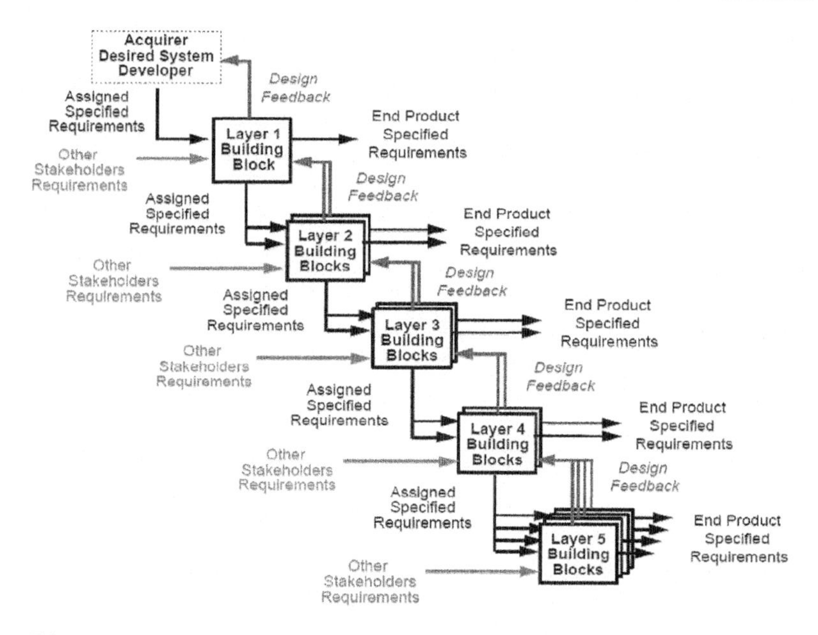

Figure 7.11. *EIA632 system design process. For a color version of this figure, see www.iste.co.uk/micouin/MBSE.zip*

Once each basic component has been produced or supplied, the system integration process will comprise the assembly of different elements of the system in a pre-established order and the verification of the conformity of successive assemblies to the intermediary specifications up to the specifications of the system itself. This is what Figure 7.11 illustrates, inspired from the EIA 632. This verification process will be described in Chapter 9.

7.8. The end of the design process and the realization

The design process ends once all the lowest level blocks have been identified and the specification models of these blocks are validated. They do not require ulterior decomposition because they are sufficiently simple components to be directly produced. This can especially be the case for software components, which can be directly coded, and electronic components, which can be directly synthesized.

A second ending condition lies in the fact that the specification of the block has been handed over to a supplier who is in charge of developing and producing it. This is what we call an acquired block.

A third ending condition to the design process lies in the fact that the physical realization of the specified block already exists and these physical realizations are consistent with the block requirements. This is block we call a COTS.

8

Validating Requirements and Assumptions

8.1. Introduction

In this chapter, we will discuss requirement and assumption validation. We recall that the signification of the word "validation" depends on the standard or norm to which it subscribes. Here, we subscribe to the ARP4754A recommendation, which is the most developed standard concerning validation; consequently, in this chapter, the concept of "validation" only refers to the concepts of requirements and assumptions, but never to the physical systems themselves.

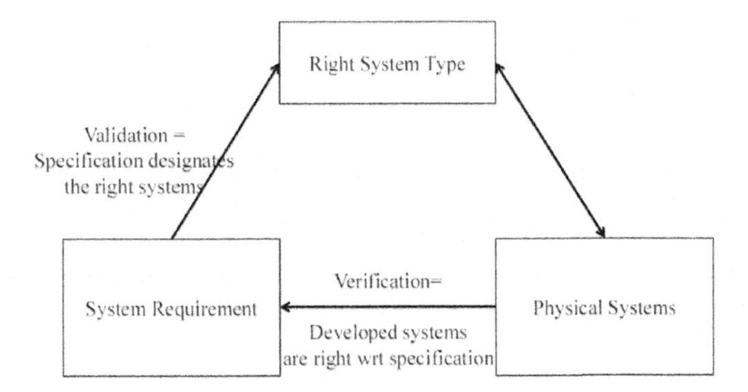

Figure 8.1. *ARP4754A validation and verification processes connection*

Therefore, validating a system does not make sense in the context of ARP4754A, and this option is fully justified by the following reasoning: to develop the right type of systems[1], it is "only necessary" that the set of all requirements characterizing it and the assumptions related to its environment be positively validated and that the systems produced with regard to its requirements and assumptions be positively verified in relation to its requirements and assumptions, that is to say compliant. The set of requirements and assumptions associated with a type of systems is the cornerstone in developing this type of systems.

First, we recall what the ARP4754A states about the validation process, as well as the goal and means that this recommendation advises and implements. We will then see how these goal and means can be declined within the framework of the systems engineering approach, based on the model we propose, that is to say the property-model methodology (PMM).

8.2. The validation process according to the ARP4754A

8.2.1. *Goal of the validation*

The ARP4754A recommendation defines validation as the process by which we ensure that the set of requirements allocated to the type of systems is sufficiently complete, and sufficiently correct, to characterize the desired type of systems, before supplying or even developing them, without taking too many risks concerning the compliance of the supplied systems with the targeted type. It also covers the assumptions made concerning its environment to specify the type of systems considered.

---

1 Right systems are those that are targeted and accepted by all the stakeholders.

We recall that the ARP4754A recommendation does not give a definition of what a requirement is, other than an evasive statement, already emphasized in this book, and it defines an assumption as a statement offered without proof [ARP XX]. It can seem strange that a body of technological knowledge will be built thus on a "missing column", according to the words of the poet Henri Michaux [MIC 29], but this case is not that exceptional. The ISO15288 has also omitted from giving a definition of the concept of requirement.

However, as formalized in section 6.2, Σ designates a targeted type of systems and $\{Req_i\}_{1\leq i\leq n}$ designates a set of the requirements referring to Σ, then the aim of the validation process is to establish that Σ is identical to $SAT_K(\{Req_i\}_{1\leq i\leq n})$. If this is not the case, a controlled modification process of $\{Req_i\}_{1\leq i\leq n}$ shall have to be executed so that we can eventually obtain $SAT_K(\{Req_i\}_{1\leq i\leq n})= \Sigma$. This is how we will have succeeded in obtaining a specification that is as exact as possible.

Thus, each requirement Req_i is individually sufficiently correct, and if $\{Req_i\}_{1\leq i\leq n}$ is sufficiently complete (it is not missing essential requirements), then we can hope that the specification$\{Req_i\}_{1\leq i\leq n}$ will also be as exact as possible.

An incorrect (respectively, an incomplete) requirement (respectively, specification) is a requirement (respectively, a specification), which can either put aside an intended characteristic of the type of systems targeted, or on the contrary conserve an undesirable characteristic to the type of systems targeted.

According to the ARP4754A, the assumptions made during the development must be explicit and follow the same validation process.

We understand, therefore, the importance of the requirement and assumption validation process, especially if we recall that these requirements constitute the cornerstone of the development process of a type of systems.

8.2.2. *Means of validation*

The approach suggested by the ARP4754A to validate a set of requirements and assumptions $\{Req_i\}_{1 \leq i \leq n}$ allocated to a type of systems involves building a validation matrix (or any other equivalent method) in which each requirement of the set $\{Req_i\}_{1 \leq i \leq n}$ is referred to by a line.

At the initial stage, for each of these lines (that is to say, for each requirement or assumption), the stakeholders in charge of the validation process select, on the one hand, a level of the rigor with which the validation process is conducted and, on the other hand, the validation method or methods to be used.

Concerning the level of rigor intended of a validation effort, the ARP4754A associates the severity of risk the crew and passengers are exposed to in case of a failure of the service provided by the system under consideration. The requirement validation level of rigor whose failure would have catastrophic consequences (CAT) for the crew and passengers must be maximal. Then, the level of rigor can be gradually reduced if the consequences are hazardous (HAZ), major (MAJ), minor (MIN) and finally no safety effect (NSE) according to a classification well established by regulation (for severity categorization, CAT, HAZ, MAJ, MIN, NSE, refer to Chapter 10, Table 10.2).

We see, therefore, that the level of rigor intended by the ARP4754A in order to conduct a validation process is entirely polarized by safety considerations, which seems normal and compliant with Authorities objectives and with what we expect in matters of aeronautic safety. However, we also see that in this framework, the requirements without any consequences on security can become the poor relation of the validation process, and it is suitable to adjust the level of rigor for requirements important for the contractor and operators such as the requirements impacting the availability, even if they have no impact of safety.

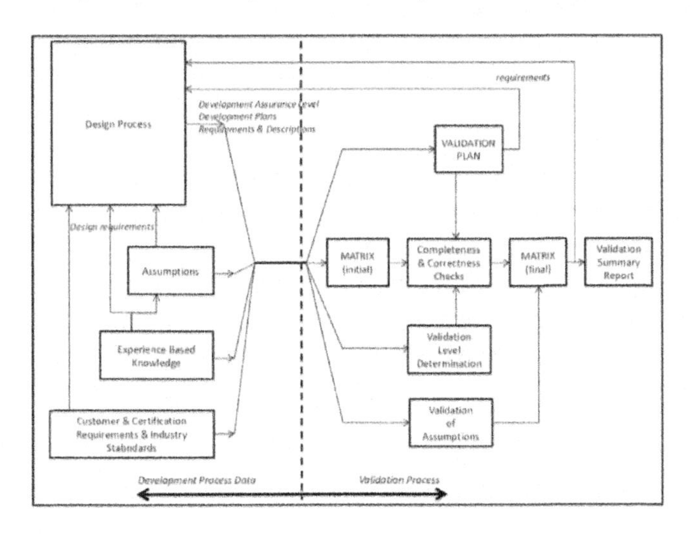

Figure 8.2. *ARP4754A validation process model*

The validation plan, which is the entrance point of the validation activities, identifies, on the one hand, the validation methods that shall be used either on their own or in a combined manner. On the other hand, it defines, depending on the intended level of rigor, the combination of methods required or advised to successfully carry out the validation process.

The ARP4754A identifies the following processes as methods of validating requirements and assumptions:

1) vertical traceability;

2) analysis;

3) modeling;

4) testing (including simulations) ;

5) similarity (operational experience);

6) engineering judgment.

These methods are used to carry out two types of checks: correctness checks and completeness checks; check-lists are established within the validation plan.

Once the initial validation matrix is established and approved, the validation process can be started, that is to say correctness and completeness checks can be carried out in compliance with the selected methods and the results from those checks recorded in the matrix. Additions, corrections or suppressions, of requirements or assumptions, shall be performed according to the rules of a controlled modification management process.

The ARP4754A recommends that this validation process should be carried out gradually in a descending manner, that is to say by starting with the highest level system (the aircraft) and then validating the aircraft system requirements, so on and so forth down to the elementary items or components. The recommendation also emphasizes that the validated system requirements are the base for the validation of requirements specified to its subsystems.

8.3. The validation process according to the property model methodology

Following the ARP4754A recommendation, the PMM introduces the requirements and assumptions model as cornerstone of the system development process. However, unlike the recommendation, which refers to no precise definition of the notion of requirement, the approach of the PMM is based on the concept of property-based requirement (PBR) introduced in Chapter 6.

8.3.1. *Goal of the validation*

To develop a type of systems, as we will see in Chapter 11, we start by developing a system specification model (SSM) of the type of systems.

This model will then be validated, that is to say, it will have to be established that this specification is as exact as possible for the type of systems targeted.

Ideally, the goal of the validation is to show that all the specification models, which constitute the specification tree, are free of specification errors.

Unfortunately, this goal is, in absolute terms, inaccessible. The claim that a specification model is free of errors can only be justified relative to the level of rigor of its validation, but remains on the existence of an error that is not detected by the validation process. However, the validation task presented here enables the reduction of this risk to acceptable levels with regard to the potential consequences.

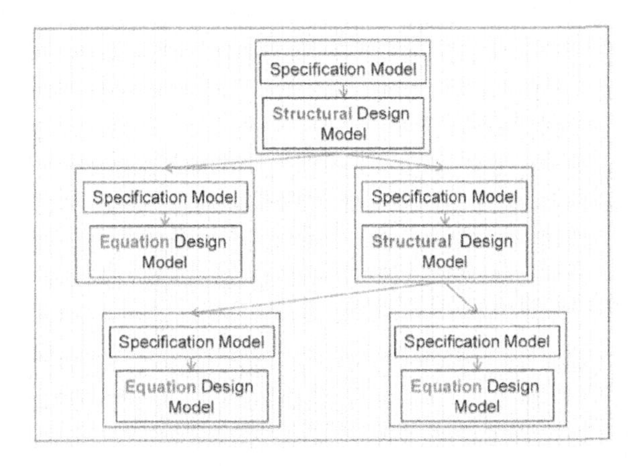

Figure 8.3. *Specification model tree validation. For a color version of this figure, see www.iste.co.uk/micouin/MBSE.zip*

Following this, an architecture of the type of systems targeted can be designed in terms of components (processes, then subsystems) and structure (flows, then links). The architectural choice enables the derivation of the SSM into a set of subsystem specification models. These subsystem specification models will then be validated in relation to the SSM they are derived from.

In other words, depending on the PMM, the specification model validation is a descending process (as suggested by the ARP4754A recommendation) which:

– starts by establishing that the specification model of the type of systems targeted is as exact as possible;

– then demonstrates that all the subsystem specification models of the tree are valid in relation to the SSM they are derived from.

This gradual requirement validation process is, through the methods indicated below, a powerful way of reducing risk and will enable us to establish an important theorem (the "contract" theorem) in Chapter 9, which we could summarize in the following way for now: if the subsystem specifications of a system are valid in relation to the system specifications and if each of the subsystems is consistent with its specification, then the system will necessarily be consistent with its own specification so long as its physical integration is compliant with its design.

8.3.2. *Means of validation*

In theory, the PMM does not put aside any of the validation methods considered by the ARP4754A (section 8.2.2). It considers them in a complementary manner (the flight test can be one, for example). However, it favors simulation as a validation method, just as it considers simulation to be the main method of cost and deadline control, and consequently the main method to reduce the complexity of a system development.

Indeed, simulation presents two advantages for validation. First, it is an *early* method of validation in comparison to methods such as tests, which are belated methods of validation[2]. In this case, it enables *a priori* the validation of the systems specification model, before the design of an architecture, whereas a validation by tests requires the availability of a prototype of the type of systems. Second, it provides an *objective* means of assessing the exactness of the specification model very superior to the early means allowed by written specifications. An SSM is not necessarily exact at the beginning of the validation process but it is objective, and this objectivity will enable us to establish whether it is exact or not, within the limits we will detail lower down. Experts (test pilots, for example) called to express themselves on the course and results of a simulation process have an objective basis to evaluate[3] the intended behavior, which more or less well-written literal specification files do not offer.

Validation by simulation involves setting up a validation workbench with, on the one hand, a system model and, on the other hand, a validation driver as shown in Figure 8.4. The validation driver is an artifact that submits validation cases, chosen from the validation scenarios, to the system whose specification we seek to validate. During each simulation cycle, the system model reacts to the validation cases submitted to it. The design model computes the outputs in function of the submitted inputs and also, more often, in function of previously determined states. Concurrently, an SSM ensures that the inputs are in

2 As they are belated in the sequence of tasks, when the time comes for tests, most of the initial budget has already been used, leaving no room for change.

3 Equivalent to those offered later by laboratory tests.

accordance with the expressed assumptions and the outputs (and observable states) are in accordance with the specified requirements.

The topics related to validation scenarios and the validation effort required will be tackled in section 8.3.5.

8.3.3. *Exactness of a system specification model*

If the design model is an equation design model (EDM), as defined in section 7.4.2, then this behavioral model is an "embodiment" of the specification model. It behaves objectively and ideally in relation to the specification, and no discrepancy can be observed between the behavior that is specified and intended, and the effective model behavior (see section 2.3). It is, therefore, possible to judge the validity of the specification by observing the effective behavior of the EDM.

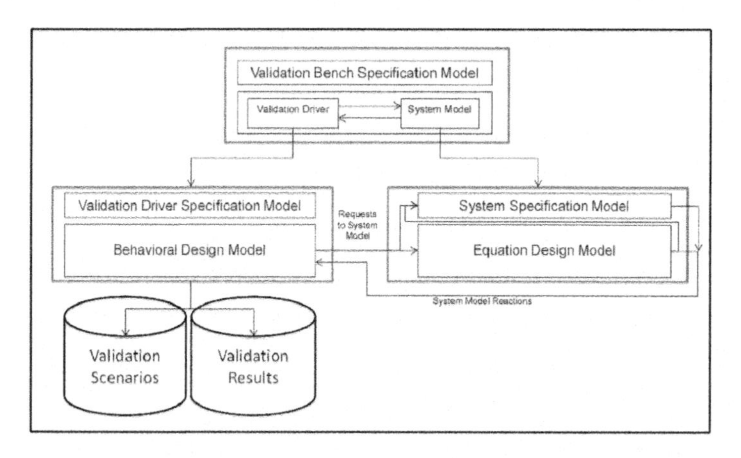

Figure 8.4. *System specification model exactification process by simulation*

If the effective behavior of the EDM satisfies the stakeholders in charge of the validation of the specification model (SSM), we can conclude that the specification model exactly characterizes the type of systems targeted.

However, the claim that the specification model is exact remains limited and approximate. This limitation is due to the fact that a model, as representative as it may be of a concrete object, cannot entirely characterize it (the real object exceeds its models in all respects), rather than due to the fact that it is conditioned by the wealth of validation scenarios, and the operational representativeness of these scenarios (see section 8.3.5). The more a specification model is validated through a large number of validation scenarios, the more exact it will be, without ever being able to attain an absolute exactness.

8.3.4. *Validating the derivation of system requirements*

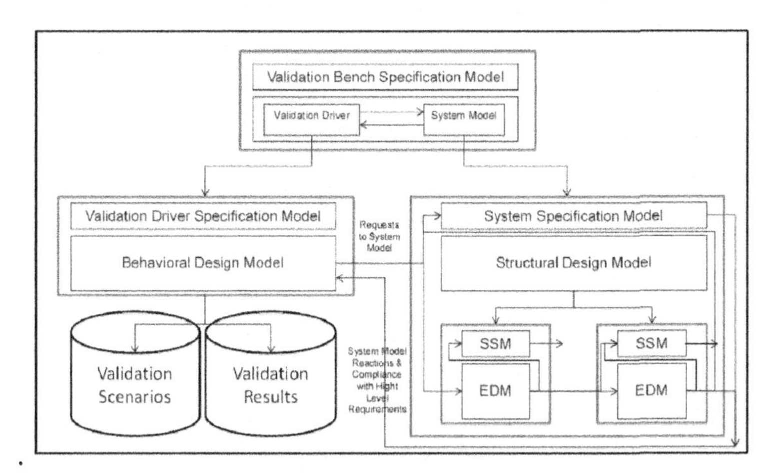

Figure 8.5. *Subsystem specification model validation process by simulation*

Just as the ARP4754A recommends, the validation process implemented here is gradual. Once the exactness of the SSM is established at the highest level, we will stick to demonstrating that the requirement specified to the first rank subsystems correctly and completely derives the system requirements.

This demonstration is carried out by a simulation, which is based on validation scenarios that are defined with regard to the required validation effort. We note, however, that this validation cannot be carried out individually, requirement-by-requirement, unlike what the validation matrix (validated line-by-line) could suggest, but requires a global awareness of each derivation operation.

In fact, we do not directly validate the requirements derived and allocated to the subsystem models, but rather the derivation task carried out in a context defined by the structural design model (SDM), which interconnects the subsystem models. It is just a misuse of language to declare that derived requirements are valid when the derivation is validated. Furthermore, we will note that an SDM is not subject to validation.

8.3.5. *Scenarios and validation cases, efforts and rigor in validation*

As we have already indicated, a specification model can be represented graphically as shown in Figure 8.6.

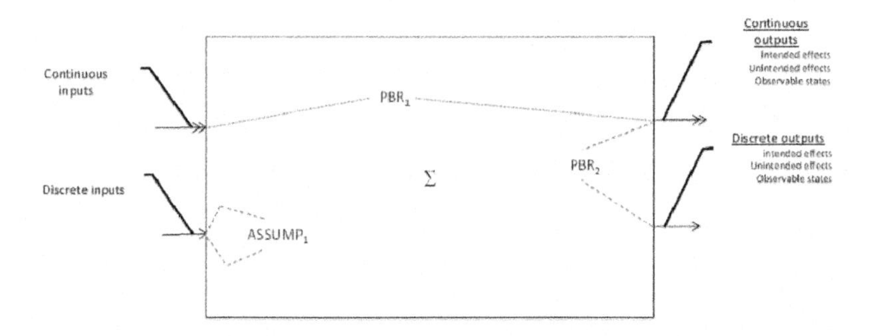

Figure 8.6. *Graphical representation of a specification model*

Each of the PBRs presented above is an expression in the form of PBR: [when C] $\rightarrow$ val(O.P) $\in$ D. Such an expression is logically equivalent to the assertion [val(O.P) $\in$ D] or not (C), or even to the expression O.P(C,D) where O.P(C,D) is a Boolean function P: $E_1xE_2x..xE_n \rightarrow$\{false, true\}, and so the set of PBRs specifying the type of systems is a set of Boolean functions with parameters selected from the inputs or outputs (or observable states) of the SSM. E_i for $1 \leq i \leq n$ is the definition domain (still named: the type) of an input, output or observable state belonging to the SSM.

During a simulation process, assertions $P(x_1,x_2,..,x_n)$ are evaluated simultaneously, that is to say that at the start of each simulation cycle, all those that require reevaluating (because the values of certain of their pertinent parameters[4] $(x_1,x_2,..,x_n)$ are changed) are reevaluated "in parallel". Under these conditions, one or more of the reevaluated assertions can become untrue, thus reporting the non-respect of requirements.

We name a validation case $(x_1,x_2,..,x_n)$ any element of the definition domain of a requirement P: $E=E_1xE_2x..xE_n \rightarrow$\{false, true\}. Depending on the types involved in the definition domain $E_1xE_2x..xE_n$, there is potentially an infinite number of validation cases. An exhaustive simulated validation of a specification model is, therefore, unattainable. This leads to the partitioning of the definition domain P into equivalence classes $P/\mathcal{R}$ such that all the validation cases belonging to the same class will be supposed to be equivalent (i.e. having the same truth value). It is then enough to choose just one validation case within each class to have a

4 All the parameters are not necessarily pertinent; for example, the change of an input must not necessarily automatically launch a reevaluation of P that can only be reevaluated on an output after the propagation of the change in the system model. The specifier determines the moments during which a requirement must be evaluated.

systematic validation strategy and a finite and representative set of validation cases at our disposal.

Let us use the following example regarding the indication of the barometric altitude on an aircraft (airworthiness requirement CS2x 1303,b). This requirement states that the altimeter of an aircraft must give an indication of the barometric altitude (ADC.Indicated Alt) with an error inferior or equal to 25 ft when the aircraft is at a real altitude (AC.Alt) between 0 and 5,000 ft, which is translated by the following PBR:

PBR_1: when *AC.Alt ∈[0.0, 5,000.0] →|ADC.Indicated_Alt - AC.Alt | ≤ 25.0 ft*

The rest of the domain [5,000.0 and 50,000.0 ft] is covered by the PBR_2 up to PBR_9 and PBR_0 requirements stated in section 6.6.1.

This PBR_1 requirement is translated, within the PMM methodology, by a Boolean function P: [0.0, 5,000.0]xR →{false, true} where R is the real type (field of real numbers) and [0.0, 5,000.0] is the interval of real values between 0.0 and 5,000.0 and formally defined by the predicate P: ∀ Alt ∈ [0.0, 5,000.0], ∀ Indicated_Alt ∈ R, P(Alt, Indicated_Alt) =true if | Alt - Indicated_Alt| ≤25.0; otherwise, P(Alt, Indicated_Alt)= false.

The exhaustive validation of this Boolean function P is impossible within a simulation process; however, the validation cases described in Table 8.1 can provide an indication of the exactness of the Boolean function.

#	Validation cases		Oracle (intended results)
	Alt	Indicated_Alt	
1	2,500.0	2,480.0	True
2	2,500.0	2,450.0	False

Table 8.1. *Validation case for a level of rigor consistent with no safety effect (NSE)*

If we use this Boolean function in simulation by injecting the validation cases above as parameters, and if we obtain the intended result in each case, the corresponding requirement could be presumed to be exact.

We can see, considering the circumstances, the lack of evidence produced to claim the requirement is exact. This lax attitude toward the validation effort could be acceptable if the erroneous altitude indication had NSEs. In fact, for the Altitude Indication, this is not the case. Consequently, we must harden the demonstration of the exactness of the requirement by increasing the quantity of evidence produced, in a reasoned and systematic manner.

It is not within the scope of this work to set the level of rigor and the quantity of evidences to produce to be able to declare a requirement to be as exact as possible. This responsibility belongs to the developing team, and it must fall within the scope of their validation plan. However, we can indicate that as the Boolean functions express themselves as sequential algorithms in simulation language, all the software-testing techniques are available to select the validation cases, pertinently and systematically. We can, for example, refer ourselves to Chapter 4 of the software engineering body of knowledge (SWEBOK V3.0) [BOU 14] dedicated to software testing.

We can, for example, harden the demonstration of requirement validation by considering the set of validation cases of the table above.

We could of course add many others, but we must ask ourselves the question of the benefit expected from a multiplication of the validation cases. From this point of view, software-testing technologies provide objective selection criteria.

#	Validation cases		Oracle (intended results)
	Alt	Indicated_Alt	
1	0.0	–26.0	False
2	0.0	–24.0	True
3	0.0	24.0	True
4	0.0	26.0	False
5	2,500.0	2,480.0	True
6	2,500.0	2,470	False
7	2,500.0	2,524.0	True
8	2,500.0	2,526.0	False
9	5,000.0	4,974.0	False
10	5,000.0	4,976.0	True
11	5,000.0	5,024.0	True
12	5,000.0	5,026.0	False

Table 8.2. *Validation case for a hardened level of rigor*

Once the validation cases enabling the exactification of a requirement are selected, we will have to inject these cases into one or more simulation scenarios, corresponding, if possible, to system usage scenarios (operational scenarios). Injecting these validation cases into operational scenarios enables the detection of situations insufficiently covered by the validation cases, and therefore the correction of this situation by enhancing the coverage of some critical phases.

Phase	Validation case #	Oracle
Taxiing-rolling	1	False
	3	True
Takeoff and climbing	5	True
	7	True
	9	False
Level flight	10	True
	12	False
Descent and landing	11	True
	8	False
Taxiing-rolling	4	False
	2	True

Table 8.3. *Validation scenarios*

We can see, in this example, that the takeoff and landing situations, which are the most delicate, are not covered much and that there might be benefits in covering them more.

8.4. Conclusion

The property-model methodology enables the development of a requirements and assumptions validation strategy that is entirely in accordance with the ARP4754A recommendation.

It enables the validation of all the specification models of a type of systems from the highest level to the most detailed levels, in a descending manner by ensuring, on the one hand, the exactness of the head specification and, on the other, by establishing the coherence, correction and completeness of the inferior level specification models in relation to the highest level specification model.

A simulation process establishes the evidence of this early validity, well before the components of the corresponding physical systems are designed and then produced.

The quantity of proofs can be of varying size and the rigor in the demonstration can be more or less hardened, depending on the stakes, particularly safety, carried by the type of systems targeted, while the validation coverage is determined systematically and adaptively to usage models.

9

Verifying the Implementation Step by Step

9.1. Introduction

In this chapter, we will discuss the verification carried out at different stages of the realization (implementation) of a given system. We recall that the signification of the term "verification" depends on the standard or the norm with which it is used. Here, we work within the framework of the ARP4754A recommendation as we did for the validation, and in order to remain consistent, within a validation/verification cut.

Consequently, the concept of verification only refers to the different stages of the implementation of a given type of systems, and there is no sense in discussing the verification of a software specification in relation to the system specification it depends on, as in the case of the DO-178C [EUR 12].

First, we will recall what the ARP4754A states about the verification process, as well as its goal and the means that this recommendation encourages and applies. Second, we will then see how these goal and means can be declined within the framework of the systems engineering approach based on the models we propose, namely the Property Model Methodology (PMM).

9.2. The verification process according to the ARP4754A

9.2.1. *Goal of the verification*

The ARP4754A recommendation defines the verification as a process by which we ensure that all the levels of implementation of a given type of systems are consistent with the requirements specified to them and how, once integrated and installed, such systems are consistent with their own specification. Its goal is to establish that, for a given level of verification rigor, related to the severity of associated failure consequences, the implementation of a system is free of errors. Verification goal and validation goal are complementary, and the conjunction of both should ensure that the implemented and installed systems are the right systems[1], as shown in Figure 8.1.

9.2.2. *Verification methods*

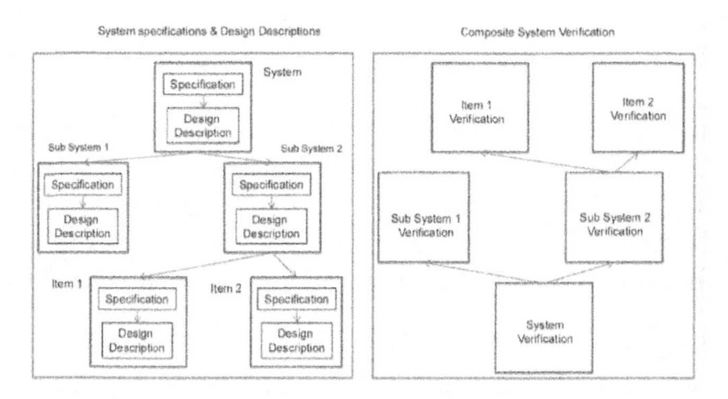

Figure 9.1. *Implementation verification process regarding specifications*

1 Right systems are those that are targeted and accepted by all the stakeholders.

Conversely to the validation process, the ARP4754A recommends an ascending verification process by first verifying the compliance of the simplest components with their own requirements and then by verifying that the integration of these components is compliant with the corresponding requirements, and by gradually proceeding thus, up to the highest system level.

During these verification processes, the non-compliance of an implementation with its specifications, due to an implementation error, can be detected. This discrepancy should then be treated according to the rules of a controlled process managing corrections and modifications.

The approach proposed by the ARP4754A to verify an implementation regarding the requirements $\{Req_i\}_{1 \leq i \leq n}$ allocated to it involves the construction of a verification matrix (or any other equivalent method) in which each requirement is referred to by a line.

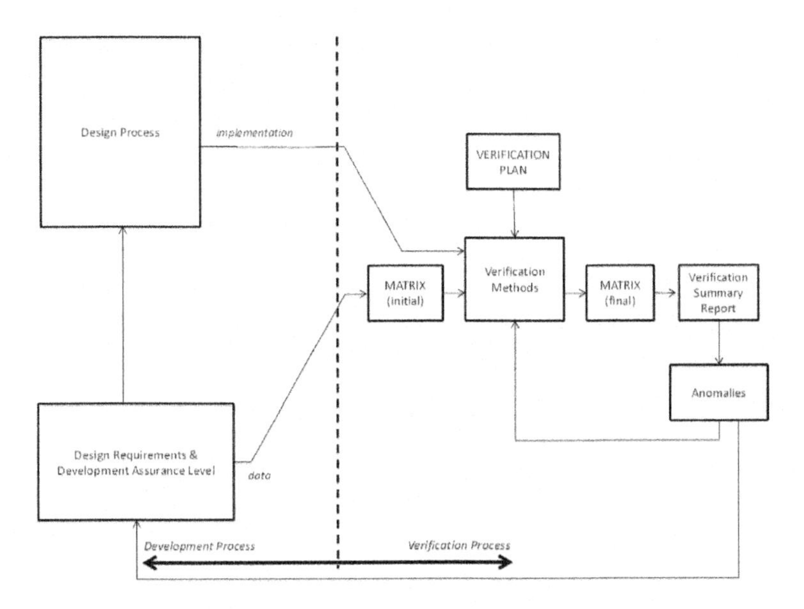

Figure 9.2. *ARP4754A verification process model*

At the initial stage, for each of these lines (that is to say, for each requirement and each assumption) the stakeholders in charge of conducting the verification process propose, on the one hand, a level of rigor with which the verification process will be conducted and, on the other hand, the verification method or methods to be used.

Concerning the level of rigor expected of a verification effort, the ARP4754A associates (just as for the rigor in validation) the severity of risk the crew and passengers are exposed to, in case of a failure of the service provided by the system under consideration. The level of rigor of the implementation verification with regard to requirement allocated to a system whose failure would have catastrophic consequences (CAT) for the crew and passengers shall be maximal. Then the level of rigor can be gradually reduced if the consequences are hazardous (HAZ), major (MAJ), minor (MIN) and, finally, no safety effect (NSE) according to a classification well established by regulation.

The verification plan, which is the entrance point of the verification activities, identifies, on the one hand, the verification methods that will be used either on their own or in a combined manner. On the other hand, it defines, depending on the intended level of rigor, the combination of methods required or advised to successfully carry out the verification process.

The ARP4754A identifies the following verification methods (implementations in relation to the requirements allocated to them):

1) inspection;

2) analysis, including verification cover analysis;

3) tests or demonstrations;

4) similarity (service experience).

We will note that the ARP4754A does not clearly state the nature of the design products, whether it should be architectural drawings or descriptions of the intended behavior; the expression "design verification", present in the ARP4754 version, has disappeared from the ARP4754A.

9.3. The verification process according to the property model methodology

9.3.1. *Objects to be verified*

Like the ARP4754A, the PMM introduces a verification process, which is applied to the implementation. This verification process covers, therefore, all the implementation phases and all the products resulting from these phases. This verification is of course carried out in relation to the corresponding system specification models.

The implementation phases include:

1) design (architectures, design of operational and failure modes and design of intended behaviors, design of failure prevention means, and so on);

2) production of unit components;

3) integration of assemblies, hardware, items, subsystems and systems;

4) installation of the system in its environment.

Design (descriptive models) or physical products are matched to the implementation phases and undergo a verification process in relation to the corresponding specification models:

1) verification of the design, namely the design models;

2) verification of the production, namely the unit components produced;

3) verification of the integration, namely the objects that are integrated up to a system of the type considered;

4) verification of the installation, namely a system of the type under consideration, installed in its environment.

9.3.2. *Goal of the verification*

Like the ARP4754A, when using the PMM, the goal of the verification is to ensure that the products of each stage of the implementation are compliant with the requirements they support.

Ideally, the verification could have aimed to demonstrate that for the following levels:

1) design models do not present any design errors;

2) unit components produced do not present any production errors;

3) the integration into subsystems and then systems does not present any integration errors;

4) the installation of a system in its environment does not present any installation error.

Unfortunately, this absence of errors at each implementation stage cannot generally be established in absolute terms. It can only be a more or less corroborated conjecture depending on the rigor of the verification process used. However, an implementation always remains at the mercy of a non-detected error because none of the verifications carried out have uncovered it. Nevertheless, the verification approach presented here enables the reduction of this risk to an acceptable level, with regard to the potential consequences of such an error.

We also see that errors can be introduced in each implementation phase of a system of the type considered, which ruins all hope of removing this or that verification phase. Such a removal would only result in a displacement of the problems such as the discovery of a design error during the verification of the installation.

9.3.3. *Verifying the design*

In this section, we will focus on the verification of the design models; we will discuss on other implementation phases in section 9.3.4.

Let us consider a composite system model such as the one presented here.

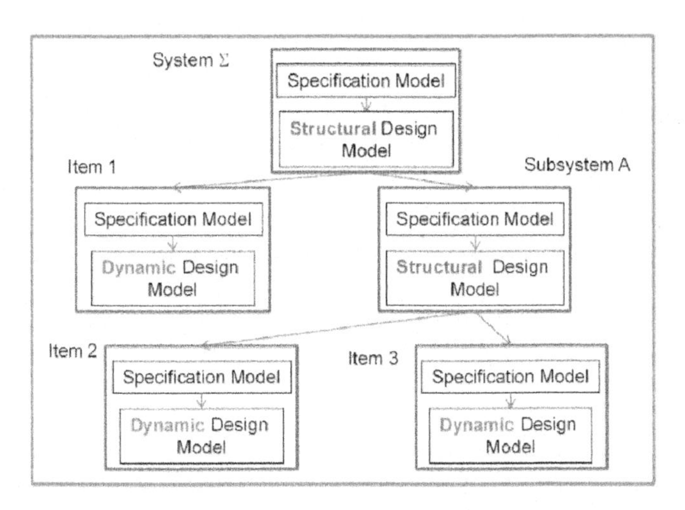

Figure 9.3. *Composite system model. For a color version of this figure, see www.iste.co.uk/micouin/MBSE.zip*

We suppose that all the specification models have been validated, with the required level of validation rigor (see Chapter 8, section 8.3.3 and 8.3.4), at each level of this system model.

We will use the system in Figure 9.3 as an example; we will then carry out the verification of the design model by starting with the lowest level objects:

1) Thus, we first carry out the unit verification of the design models of items 1, 2 and 3.

2) We then verify the integration of the subsystem design model A.

3) We finally verify the integration of the design model to the system Σ.

4) When appropriate, we verify the installation of the system design model in a higher level model that includes Σ's environment.

PMM does not put aside any methods of verification considered by the ARP4754A (section 9.2.2). However, it favors simulation as a verification method at design level, just as it considers simulation to be the main method of cost and deadline control, and consequently the main method to reduce the complexity of a system development.

By carrying out a design verification by simulation, we have the opportunity to detect the design errors early, which is always preferable.

Simulation presents two advantages for design verification. First, it is an early method of verification in comparison to methods, such as tests, which are belated methods of verification[2]. Second, it offers an objective method to evaluate the design model, whereas other early methods based on literary descriptions (such as analyses and inspections) depend more on the people carrying them out. Initially, a system design model can contain design errors,

2 As they are belated in the sequence of tasks, when the time comes for tests, most of the initial budget has already been used, leaving no room for change.

but it is objective and this objectivity will enable us to consider it free of errors, with a sufficient degree of confidence, after having carried out a verification effort proportional to the potential risks.

Verification by simulation implies setting up a verification workbench involving, on the one hand, a system model and, on the other hand, a verification driver as shown in Figure 9.4.

The verification driver is an artificat that submits verification cases, chosen from the verification scenarios, to the systems model, which we want to verify the design. The model reacts to the verification cases submitted to it at each simulation cycle. The design model computes its outputs as a function of the inputs submitted to it and also, more often, as a function of observable states. During that time, the specification model ensures that the inputs are compliant with the assumptions made and the outputs (and observable states) are compliant with the specified requirements.

The question of verification scenarios and the verification effort required will be discussed below.

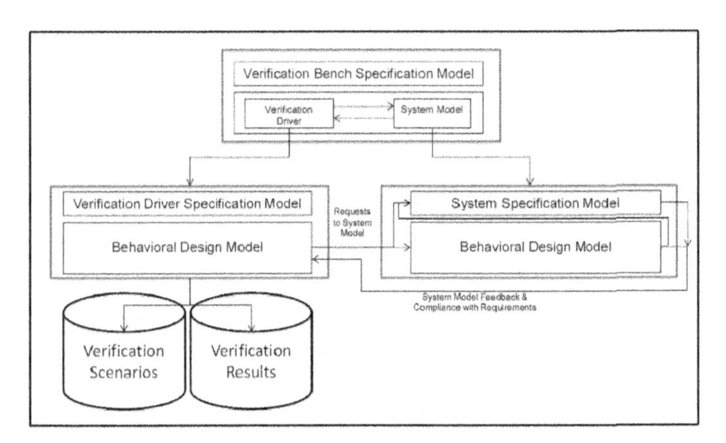

Figure 9.4. *System model verification bench*

When a design model is not an equation design model (EDM) but a behavioral design model (BDM), we can legitimately expect the specification model to detect non-compliances in relation to the requirements (that is to say, design errors), when the verification scenarios are submitted to the system model. It is appropriate to treat these discrepancies according to the rules of a controlled correction and modification management process, until the specification model stops detecting discrepancies.

Furthermore, if the verification scenarios comply with the effort of verification rigor intended, and that no non-compliances are detected, the design model can then be considered as free of design errors.

The system design model being verified is generally a composite model whose most detailed elements are BDMs like the one presented in Figure 9.5.

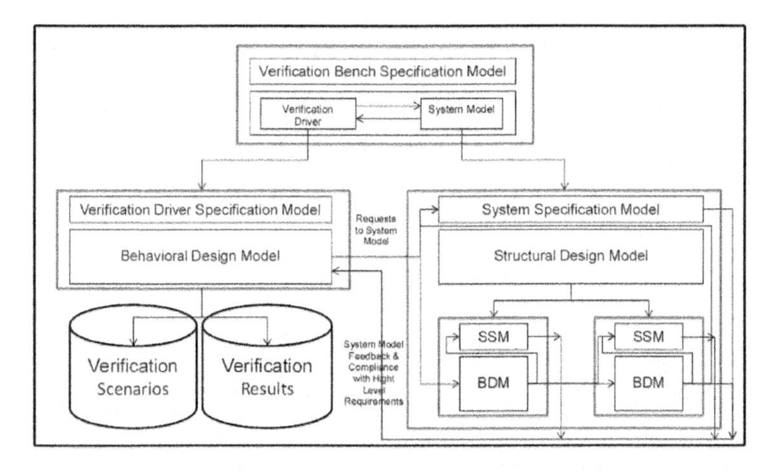

Figure 9.5. *System model integration verification bench*

If we adopt the gradual and ascending approach to verification, which is presented above, the lowest level BDMs are verified first until they can be considered compliant with their respective specification models. Once this step is complete, the integrated models can be verified.

Furthermore, if the verification scenarios comply with the effort of verification rigor intended and no non-compliances are detected, then the design model can gradually be considered free of integration design errors up to the global system level.

Considerations related to the rigor and validation effort mentioned in section 8.3.5 of the previous chapter can be completely transposed to the design verification, to the verification effort and its rigor. Then, it is clear that the scenario definition strategies and verification cases of the different dynamic design models can be completely traced from those of scenarios and the validation cases.

Furthermore, we can, if we wish, integrally reuse the scenarios and validation cases, as scenarios and verification cases. If we wish, we can also complete them.

9.3.4. *Verifying the other products of implementation*

Unlike the design models, which are semiotic objects, the other products of implementation are material objects, mechanical parts, electro mechanic machines, chemical reactors, electronic components, hardware loaded with their software, electromagnetic communication systems, etc. They are the result of production, integration and installation processes, which can all be faulty.

Once each elementary component has been produced or acquired, it is verified. Then, the system integration process will involve assembling the different system elements in a pre-established order and verifying the compliance of the successive assemblies with the intermediary specifications up to the specifications of the system itself. This is illustrated in Figure 9.6, inspired from the EIA 632.

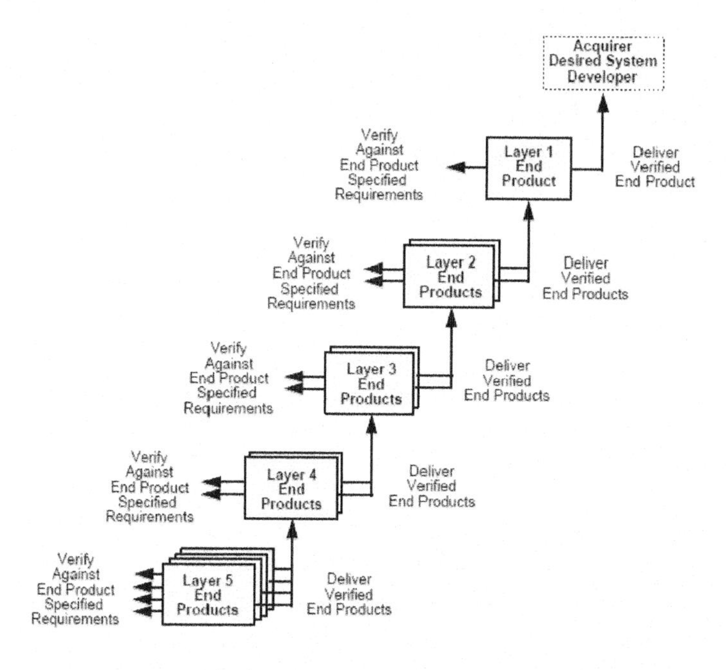

Figure 9.6. *EIA632-based system verification process*

The possibility of an error, being inscribed in the material product at any phase of the implementation stages, imposes a verification, which is not carried out using models and simulation methods, but on material products by testing.

The test bench and the aircraft now replace the simulation workbench, and the hardware replaces the models. However, subject to possible adaptations to differences in support, the set of simulation scenarios and verification cases can be reused. This is especially true as the compliance with certain requirements can only be verified in the real world, such as calibration requirements and the worst-case execution time (WCET).

9.3.5. *The contract theorem*

This enables us to conclude with a fundamental theorem of systems engineering, which we could name the "contract theorem".

Let Σ be a type of systems, defined by a specification model $\{Rq_i\}_{1\leq i\leq n}$, where each Rq_i is a property-based requirement (including assumptions).

Suppose an architecture A, selected for Σ, is constituted of m subsystem types $\{ \sigma_1, .., \sigma_m\}$ linked between each other following the endo-structure defined by A and that each requirement Rq_i allocated to Σ is derived over different subsystems σ_j of Σ such that:

when $A \rightarrow Rq_i \leq Rq_{i,1} \wedge .. \wedge Rq_{i,m}$ (i), for all i belonging to [1, n],

We can then establish the following theorem:

Premises

1) if $\{Rq_i\}_{1\leq i\leq n}$ is an exact specification of Σ;

2) if, for all i of [1, n], the derivation of Rq_i according to A is validated;

3) if the integration of the physical subsystems $\{s_j\}_{1\leq j\leq m}$ as a physical system S of type Σ is consistent with A;

4) if, for all j of [1, m], the physical subsystem s_i of the σ_j type is compliant with its specification model $\{Rq_{i,j}\}_{1\leq i\leq n}$;

Conclusion

5) then, the physical system S of type Σ is a right system.

9.4. Conclusion

The PMM enables the development of a verification strategy of the different stages of the implementation in relation to requirements fully compliant with the ARP4754A recommendation.

It enables the verification of all the design models of a systems model, ascending from the most detailed to the highest levels, and ensuring the compliance of the system design with the specification model, step-by-step, with a degree of confidence proportional to the rigor of verification effort developed.

The evidence of this design verification is established early by simulation, well before the components of the corresponding physical systems are realized.

It then enables the verification of all the production, integration and installation artifacts, ascending from the most elementary to the highest level, ensuring the compliance of the system implementation with its specification, step-by-step, with a degree of confidence proportional to the rigor of verification effort developed.

10

Safety Engineering

10.1. Introduction

Most states[1] impose that a level of protection should be provided by civil aviation to each citizen through the adoption of safety rules, and by measures that ensure that products, people and organizations respect these rules. This is the reason why aeronautical products[2] are subject to certification to guarantee that they comply with the essential airworthiness requirements related to civil aviation, airworthiness being the aptitude of a civil aircraft to safely carry out its mission: transporting people.

In this chapter, we will discuss safety engineering carried out during the different phases of the development of a type of systems. We recall the signification of the term "safety" given by the ED79A/ARP4754A at page 9: safety is a state in which the risk is acceptable, whereas the term "risk" (p. 9)

---

1 As an example, Regulation (EC) No 216/2008 of the European Parliament and of the Council of 20 February 2008 on common rules in the field of civil aviation and establishing a European Aviation Safety Agency, and repealing Council Directive 91/670/EEC, Regulation (EC) No 1592/2002 and Directive and repealing Council Directive 91/670/EEC.

2 "Aeronautical product" means any aircraft, aircraft engine, aircraft propeller or aircraft appliance or part or the component parts of any of those things, including any computer system and software.

corresponds to the probability of occurrence of an event associated with its severity.

First, we recall what the ARP4754A says of the safety assessment process, the goal that this recommendation sets and the means that it advocates. We will then see how these goals and means can be declined within the framework of the systems engineering approach based on the models we propose, namely the property-model method.

10.2. The safety assessment process according to the ARP4754A

10.2.1. *Goal of safety assessment process*

Strictly speaking, the ARP 4754A recommendation defines the safety assessment as the process which, carried out jointly with other ARP4754A processes, enables us to ensure the compliance of an aircraft system with airworthiness requirements such as the 1309 requirement of the US Federal Aviation Regulations (FARs) or the EASA Certification Specifications (CSs). It relies on interpretative material produced by the regulatory authorities (e.g. Federal Aviation Administration (FAA) and EASA) such as the advisory circular (AC) 1309 in the US or then acceptable means of compliance (AMC) 1309 in Europe, and called "System Design and Analysis".

Broadly, safety assessment is considered as a process with which we demonstrate the safety requirements, posited by the law, have correctly and completely been taken into account during the development of aircraft systems dedicated to civil aviation for the transport of persons and cargo.

Safety assessment is a process that establishes that the safety of persons (crew and passengers) is ensured in specified operation conditions and specified maintenance conditions of the system whose installation is certified, going from the identification of failure conditions of functions ensured by the system considered to the demonstration that the risks remain acceptable.

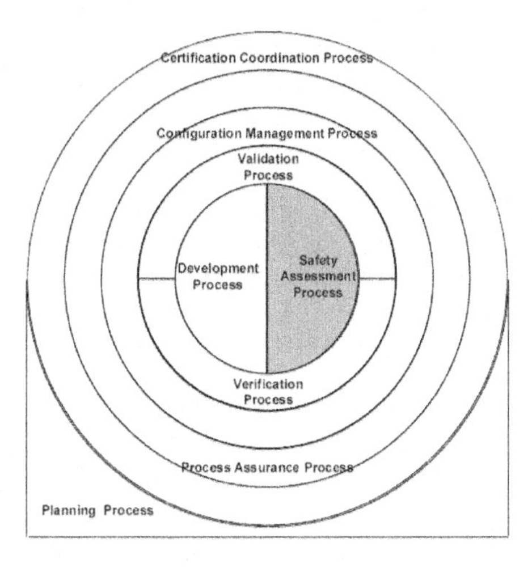

Figure 10.1. *ARP4754A safety assessment process*

10.2.2. *Means to assess safety*

First, the ARP4754A endorses the definitions provided by the interpretative material issued by authorities.

This interpretative material (for example, AC29.1309.b(1)) quantifies the propensity of failure conditions according to a scale of failure rate per flight hour, established as follows:

Propensity classification	Range lower bound	Range upper bound
Frequent		10^{-3}/fh
Reasonably probable	10^{-3}/fh	10^{-5}/fh
Remote	10^{-5}/fh	10^{-7}/fh
Extremely remote	10^{-7}/fh	10^{-9}/fh
Extremely improbable	10^{-9}/fh	

Table 10.1. *Failure rate classification according to AC29.1309.b(1)*

It also sets (for example, AC29.1309.b(2)) failure condition severity quantification criteria, ranging from "No Safety Effect" (NSE) to "Catastrophic" (CAT) and passing by "Minor" (MIN), "Major" (MAJ) and "Hazardous" (HAZ) according to a severity scale defined in Table 10.2.

Severity	Definition
No effect (NSE)	Failure conditions that would have no effect on safety
Minor (MIN)	Failure conditions that would not significantly reduce rotorcraft safety and that would involve crew actions that are well within their capabilities
Major (MAJ)	Failure conditions that would reduce the capability of the rotorcraft or the ability of the crew to cope with adverse operating conditions to the extent that there would be, for example, a significant reduction in safety margins or functional capabilities, a significant increase in crew work load or in conditions impairing crew efficiency, physical distress to occupants, possibly including injuries, or physical discomfort to the flight crew
Hazardous (HAZ)	– a large reduction in safety margins or functional capabilities; – physical distress or excessive workload such that the flight crew's ability is impaired to where they could not be relied on to perform their tasks accurately or completely; or – possible serious or fatal injury to a passenger or a cabin crew member, excluding the flight crew
Catastrop hic (CAT)	Failure conditions that would result in multiple fatalities to occupants, fatalities or incapacitation to the flight crew, or result in loss of rotorcraft

Table 10.2. *Failure condition severity definitions according to AC29.1309.b(2)*

This interpretative material (AC29.1309.b(3)(ii)) then sets the acceptable levels of risk as shown in Table 10.3.

Failure condition severity	Acceptable failure condition propensity
NSE	Frequent
MIN	Reasonably probable
MAJ	Remote
HAZ	Extremely remote (improbable)
CAT	Extremely improbable

Table 10.3. *Safety objectives for installed systems according to AC29.1309.b(3)(ii)*

The interpretative material (AC29.1309.b(3)(iii)) determines also a technological operative rule according to which the safety objectives associated with CAT failure conditions are achieved if, on the one hand, (1) no simple failure can lead to a CAT failure condition, and if, on the other hand, (2) each CAT failure condition is extremely improbable (i.e. $\lambda \leq 10^{-9}$/fh).

Second, the ARP4754A extends through the concepts of functional/item development assurance level (FDAL/IDAL), concepts inherited from the ARP4754 recommendation (development assurance level (DAL)) and ED12/DO178 standards (software level) for software, and ED80/DO-254 [ED 00] (hardware design assurance level) for electronic components. A product development assurance process is a means for avoiding development errors and mitigating their consequences (in Chapter 8 (section 8.3) and Chapter 9 (section 9.3), we distinguished errors in specification, design, production, integration and installation). The rigor of this process is graduated in five levels (from the most rigorous (A) to the least rigorous (E)), and each level, *a priori*, corresponds, item by item, to the severity of the failure conditions caused by the product, as shown in Table 10.4.

Top-level failure condition severity	Top-level FDAL assignment
CAT	A
HAZ	B
MAJ	C
MIN	D
NSE	E

Table 10.4. *Top-level function FDAL assignment according to ARP4754A*

The recommendation also proposes a set of technological operational rules that allow the assignation of the FDAL/IDAL, by taking into consideration the architectural characteristics of the system or the item under consideration (namely, Table 3, pp. 41 and 42 of the ED79A/ARP4754A).

Third, it systematizes the safety assessment process sketched out by the interpretative material (for example, AC29.A309.b(4)) by detailing the different analyses that should be conducted:

– functional hazard assessment (FHA) carried out at the aircraft and aircraft system levels, concurrently to the requirement determination process;

– preliminary aircraft safety assessment (PASA) and preliminary system safety assessment (PSSA) also carried out at the aircraft and aircraft system levels, concurrently to the processes of solution design, derivation and allocation of requirements;

– aircraft safety assessment (ASA) and system safety assessment (SSA) also carried out at the aircraft and aircraft system level simultaneously to the verification process;

– common cause analysis (CCA) including common modes analysis (CMA), zonal safety analysis (ZSA) and particular risk analyses (PRA).

10.2.2.1. *Functional hazard assessment*

An FHA aims to examine the possible failures in functions (intended effects), first at the aircraft level, and then at the aircraft system level, to determine the severity of the consequences of these failures depending on the context in which they appear. The detailed manner of carrying out these FHAs is covered in the ARP4761 recommendation [SAE 96]. This last recommendation proposes (in annex A, p. 39) a format to present the results of an FHA, which are identified for:

1) the functions (of the aircraft or the aircraft system);

2) the possible failure conditions;

3) the operational phases considered;

4) the consequences of these failures on the aircraft, crew and occupiers;

5) the classification of the failure conditions according to severity;

6) the references to the justifying material;

7) the verification method enabling us to establish that the chosen design solution is compliant with the safety objectives;

8) the qualitative and quantitative safety objectives allocated to the functions of the aircraft or its systems.

10.2.2.2. *Preliminary aircraft/system safety assessment (PASA and PSSA)*

The preliminary safety assessment enables the derivation of safety objectives (or requirements) defined at the FHAs level and the assignation of derived safety requirements to the different architectural elements of the aircraft or the systems considered, namely the safety requirements assigned, on the one hand, to the structure and, on the other hand, to the components of the system. These quantitative[3] and qualitative safety requirements allocated to the system structure are design requirements aiming to reduce risks. Preliminary safety assessment goes alongside the system design process:

– *Quantitative safety requirements*: the term "quantitative", concerning the safety requirements, covers the requirements translating a maximum probability of occurrence, such as "loss of the altitude indication shall be less than or equal to 10^{-9} per flight hour".

– *Qualitative safety requirements*: conversely, the term "qualitative", concerning the safety requirements, covers the requirements such that "the system shall possess a level of redundancy greater than or equal to 3 and shall possess a level of dissimilarity greater than or equal to 2".

– *Fail safe concept*: similarly, the requirement known as the "fail safe concept", meaning that the occurrence of a simple failure shall not have CAT consequences[4], is part of the qualitative safety requirements.

– FDAL/IDAL: the safety requirements related to DAL are also qualitative safety requirements. The detailed manner of carrying out such preliminary safety assessments is covered in the ARP4761 recommendation (annex B, p. 39).

10.2.2.3. *Aircraft / system safety assessment (ASA and SSA)*

Safety assessment aims to verify the safety objectives allocated to the functions at the FHA's level and the safety requirements allocated to the different architectural elements of the aircraft or the system considered are effectively satisfied. This safety assessment goes with the

---

3 A safety requirement is a quantitative requirement if the compliance demonstration requires numerical analytical methods (AMC25.1309, p. 2-F-41); otherwise, it is qualitative.

4 "No single failure will result in a catastrophic failure condition AC29-2C page F-18".

verification process of the system implementation. The detailed manner of carrying out such safety assessments is covered in the ARP4761 recommendation (annex C, p. 45).

10.2.2.4. *Common cause analysis*

CCAs cover three types of analyses: CMAs, ZSAs and PRAs. These analyses all aim to detect the common causes that could ruin safety arrangements and falsify safety assessments based on these arrangements. The detailed manner of carrying out such CCAs is covered in the ARP4761 recommendation (annexes I, J and K, pp. 151, 156 and 159).

10.3. The safety assessment process according to the property model methodology (PMM)

10.3.1. *Errors, faults and failures*

In Chapter 2 (section 2.6), we have introduced the concepts of failure, fault and error. These concepts have been introduced in connection to those of function, behavior and structure of the Function-Behavior-Structure (FBS) framework (see section 2.3) as presented in Figure 10.2.

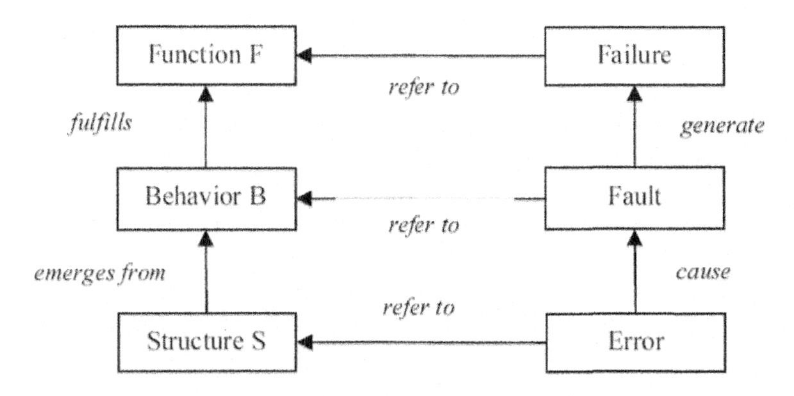

Figure 10.2. *Failure, fault and error*

Failure: when a technological system operates in environmental conditions it has been designed for, it experiences a failure the instant it ceases to produce the intended effect on its environment, whether it be the loss of the required function or the faulty performance of that function.

Faults: a faulty behavior (or symmetrically, a faulty state) of a technological system is the more or less immediate cause of a failure, i.e. a loss or faulty performance of its function.

A failure, i.e. a loss or faulty performance of its function, is the consequence of one or more faulty behaviors (or symmetrically, faulty states) of the technological system under consideration.

Errors: faulty behaviors (or symmetrically, faulty states) originate in errors either in the structure or the components of the technological system or in the assessment of the environmental conditions.

Certain faults are said to be dormant, i.e. they remain unknown from the operator during a certain period of time. For example, an engine, a pump or even a valve can be in a faulty state, but this state remains hidden as they are at a standstill. The fault only reveals itself when this engine, pump or the command of this valve are activated. The built-in tests (BITs) and preflight tests are appropriate means for the detection of dormant faults.

In a system organized in multiple architectural levels, the notions of fault and failure are relative to the level under consideration. Thus, the failures of an aircraft are caused by simple or combined faults of the aircraft systems. These faults, considered as failures at the aircraft system level, are caused by simple or combined faults of the subsystems of the system under consideration.

Finally, we will note that the use of a technological system, in environmental conditions explicitly excluded from the usage domain, cannot lead to a failure of the technological system considered but constitutes an operation error.

10.3.2. *FHA and interpretation of the 1309(b)(2)(i) requirements as PBRs*

In the different regulations, the 1309 requirement requires the aircraft and system manufacturers to install systems whose failure cannot lead to an exposure of passengers and crew to unacceptable risks.

If we consider the FAR29 and its 1309(b)(2)(i) clause

FAR 29.1309 Equipment, systems, and installations.

(b) The rotorcraft systems and associated components, considered separately and in relation to other systems, must be designed so that

1) This part of the clause is deliberatly elided this is same situation as the one encountred in the section 6.6.1. Example 1: FAR29.1303(b) flight and navigation;

2) For category A rotorcraft:

i) The occurrence of any failure condition that would prevent the continued safe flight and landing of the rotorcraft is extremely improbable.

ii) The occurrence of any other failure conditions that would reduce the capability of the rotorcraft or the ability of the crew to cope with adverse operating conditions is improbable.

In the following passage, we will show that the interpretation of this airworthiness requirement in terms of property-based requirements (PBRs) entails carrying out an FHA.

Expressed in the context of the EIA 632 requirement definition process, we will first examine the function failure conditions (failure of effects intended by the acquirer and the regulation) of the type of systems under consideration and then assess the severity of these failure conditions to set the safety requirements (top-level safety requirements), which should, if they are satisfied, render the risk acceptable.

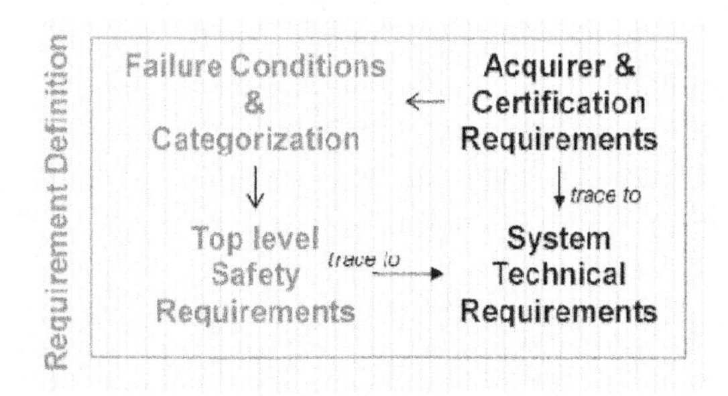

Figure 10.3. *EIA632 requirement definition process extended to safety aspects. For a color version of this figure, see www.iste.co.uk/micouin/MBSE.zip*

The 1309(b)(2)(i) statement integrates sentences such as "the occurrence of any failure condition which would prevent the continued safe flight and landing of the rotorcraft" and expressions such as "extremely improbable", which shall be interpreted according to interpretative material recalled in section 10.2.2 in the following way: i.e. catastrophic failure conditions shall have a probability of occurrence less than or equal to 10^{-9}/fh.

System or item failure conditions are specific to the system or item under consideration. Furthermore, the severity of the consequences of the failure conditions on flight and landing safety depends on (1) the system or item under consideration, (2) the system or item failure condition and (3) assumptions on the environment of the system or item (by system or item environment, we mean the other systems in the aircraft, but also the aircraft environment itself, namely the ground, the meteorological conditions and other traffic).

Figure 10.4 illustrates this process in which the airworthiness requirement 1309(b)(2)(i) is assigned to a set (*a priori,* variable) of systems installed on board an aircraft, such as an aircraft flight control system (AFCS), a fly-by-wire system (FBWS) and an air data system (ADS).

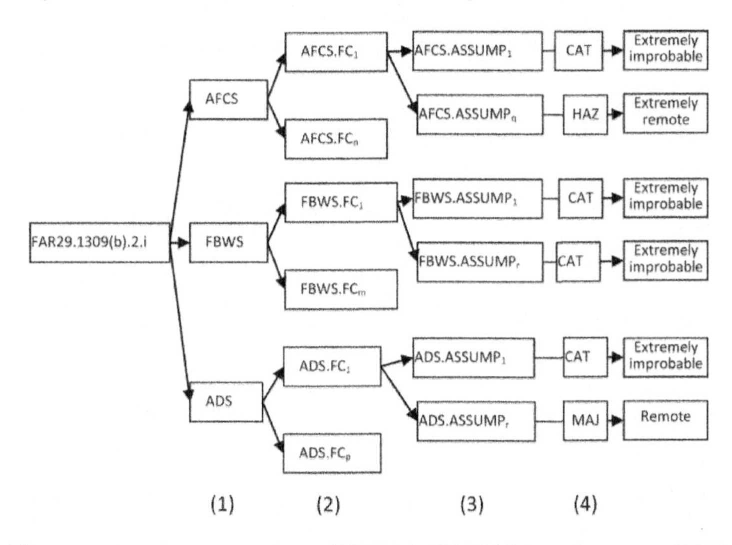

Figure 10.4. *Interpretation of FAR29.1309(b)(2) and systems FHA*

In this way, we consider, for example, an ADS, which provides an indication of the barometric altitude of the aircraft, among other air data. This indication of barometric altitude is provided to the crew and the AFCS (mainly in order to maintain a vertical separation between aircraft in the air space system) and possibly to other systems. If the aircraft evolves in instrument meteorological conditions (IMC), the provision of an undetected erroneous indication of the barometric altitude is a specific failure condition of the ADS. This failure condition is considered as CAT by the interpretative material (AC25.11A[5]). In the case considered, we can deduce the following PBR from the textual requirement FAR 29.1309.b(2):

> PBR_{10}: *when AC.Flight_Condition=IMC →*
> *val(ADS.MisleadingIndicatedAlt.propensity)* $\leq 10^{-9} / fh$

In other words, we can deduce from the previous case that the interpretation of the textual requirement FAR 29.1309.b(2) is a conjunction of PBRs of the form:

> *PBRquant: when* $Assump_i$
> *→ val(Systemi.*$FailureCondition_i$*.propensity)* $\leq \lambda_i$

where $Assump_i$ is an assumption made about the context (for example, the phase of the flight and the meteorological conditions) in which the failure condition *FailureCondition* can occur. *System* is a system likely to experience the considered failure condition *FailureCondition*, and *propensity* is the propensity of the system *System* to experience the failure condition *FailureCondition,* and λ_I is the maximum failure rate acceptable per flight hour. This enables us to specify the quantitative safety requirements assigned to the system under consideration as a conjunction of PBRs.

The interpretation of the FAR 29.1309 requirement, based on the ARP4754A, also allows us to interpret qualitative safety requirements, such as the FDAL, in terms of PBRs, assigned to the system under consideration.

> PBR_{FDAL}: *val(*$System_i$*.FDAL)=A (or B, C, D or E)*

---

5 Federal Aviation Administration, Advisory Circular AC25-11A, Electronic Flight Deck Displays, 6-21-2007.

Furthermore, the interpretative material[6] imposes additional requirements, for systems where failure conditions have CAT consequences, such as the requirement known as “fail safe concept”. This “fail safe concept” requires that a system continues producing the intended effects (its function) despite the occurrence of a single failure, which it shall tolerate.

Thus, to interpret the “fail safe concept” in terms of PBRs, the system’s intended behavior would have been previously specified in nominal conditions to then require the upholding of this intended behavior when a single failure occurs. We will illustrate this in the following example.

Consider a baro-altimeter system to be installed on board of an aircraft, which will ensure the transport of passengers according to the instrument flight rules (IFRs). In such conditions, regulation considers the loss of indication of the barometric altitude or even the indication of an undetected faulty baro-altitude as catastrophic. As imposed by the regulation, such failure conditions shall be extremely improbable, and therefore the failures rates per flight hour associated with these failure conditions shall be less than 10^{-9}. In addition, the fail safe concept applies.

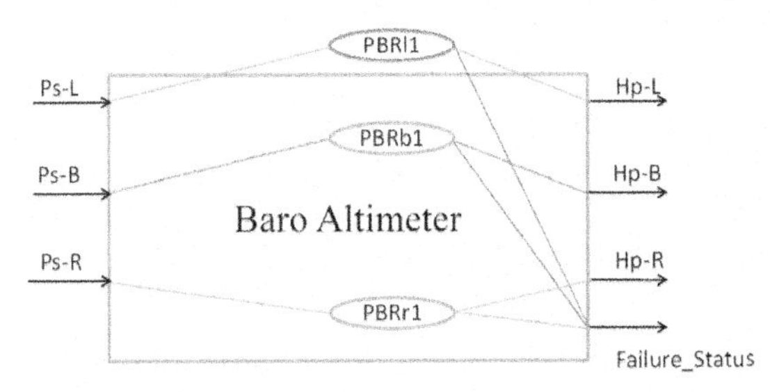

Figure 10.5. *Baro-altimeter specification model. For a color version of this figure, see www.iste.co.uk/micouin/MBSE.zip*

6 AC29.1309 in Federal Aviation Administration, Advisory Circular AC29-2C Change 3, Certification of Transport Category Rotorcraft, p. F-12, 30 September 2008.

In a first attempt, a system providing two indications of the baro-altitude devised from two pairs of static ports was considered. This first lead was quickly abandoned because it did not guarantee the non-occurrence of an undetected faulty baro-altitude indication. As a second attempt, a system providing three indications of the baro-altitude (left, back up and right) devised from three pairs of static ports was kept. The model also proposes an observable state signaling the presence, or not, of a single failure.

This model supports four PBRs: PBR_{L1} concerning the indication of left altitude, PBR_{B1} concerning the indication of the backup altitude and PBR_{R1} concerning the indication of the right altitude (where f is a real defined function), and a fourth requirement PBR_{Disp1} about the dispersion of the three indications:

PBR_{L1}:when Altimeter.Not(Failure_Status)→ HpL(t+200ms)=f(PsL(t));

PBR_{B1}:when Altimeter. Not(Failure_Status)→ HpB(t+200ms)=f(PsB(t));

PBR_{R1}:when Altimeter. Not(Failure_Status)→ HpR(t+200ms)=f(PsR(t)).

This means that we require that the value of the provided altitude indication at the left (respectively, back up, right) be calculated from the measured static pressure at the left (respectively, back up, right) port by using the function f with a maximum delay of 200 ms of the indication on the measure.

A fourth PBR is about the dispersion of provided altitude indications. It shall be less than or equal to 25 ft, in normal conditions.

PBR_{Disp1}:when Altimeter. Not(Failure_Status)→ |HpL(t- HpB(t)| ≤25ft and |HpB(t- HpR(t)| ≤25ft and |HpR(t- HpL(t)| ≤25ft

From there, the “fail safe concept” imposes that this system continues to produce the intended effects (that is to say it provides the three indications of the baro-altitude: left, back up and right) despite the occurrence of a single failure, which it shall tolerate.

In these conditions, the single failure tolerance supposes an extension of the specification model by adding three additional requirements, PBR_{L2} for the left indication, PBR_{B2} for the backup indication and PBR_{R2} for the right indication:

PBR_{L2}*:when Altimeter.SingleFailure* $\rightarrow$
$HpL(t+200ms)=f(PsL(t)) or f(PsB(t))$;

PBR_{B2}*:when Altimeter. SingleFailure* $\rightarrow$
$HpB(t+200ms)=f(PsB(t))$ or $f(PsR(t))$;

PBR_{R2}*:when Altimeter. SingleFailure* $\rightarrow$
$HpR(t+200ms)=f(PsR(t))$ or $f(PsL(t))$.

In the same way, the dispersion requirement is extended to a single failure condition.

PBR_{Disp2}*:when Altimeter. SingleFailure* $\rightarrow$
$|HpL(t- HpB(t)| \leq 25ft$ and $|HpB(t- HpR(t)| \leq 25ft$ and
$|HpR(t- HpL(t)| \leq 25ft$;

The requirements PBR_{L2}, PBR_{B2} and PBR_{R2} are, respectively, less constraining than the requirements PBR_{L1}, PBR_{B1} and PBR_{R1}. Indeed, when a single failure occurs (we do not know which one), they allow the indications H_{pL} (respectively, H_{pB}, H_{pR}) to be computed with the same constraints, not from a unique source (respectively, P_{sL}, P_{sB}, P_{sR}), but from one or the other of the two sources P_{sL} or P_{sB} (respectively, P_{sB}, or P_{sR}, P_{sR} or P_{sL}). In other words, the tolerance to single failure will be carried out through a source diversification. In the same time, results dispersion has to be held (PBR_{Disp2} instead of PBR_{Disp1}).

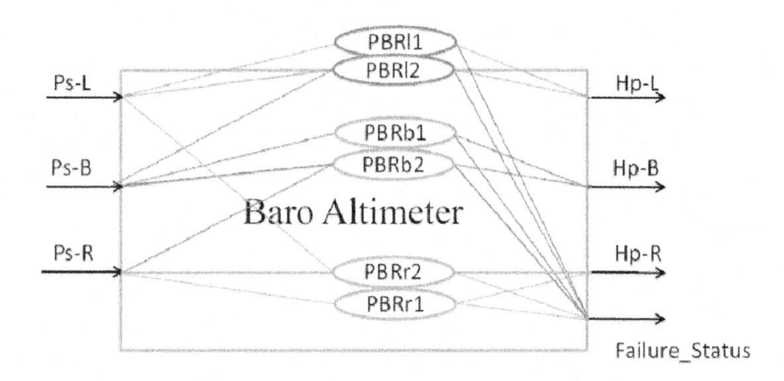

Figure 10.6. *Fault-tolerant baro-altimeter specification model. For a color version of this figure, see www.iste.co.uk/micouin/MBSE.zip*

10.3.3. *PASA/PSSA and deriving safety requirements*

As we saw previously, the safety requirements, be they quantitative or qualitative, are outcomes of the FHA of the aircraft/system functions and will be assigned to components and aircraft/system structure in the context of a PASA/PSSA, in the way it is synthesized in the ARP4754A and developed in the ARP4761 recommendation annex B.

In the following section, we will show that the derivation of safety requirements (quantitative or qualitative) with respect to a given design architecture amounts to leading a preliminary safety assessment (PASA/PSSA).

Expressed in the context of the EIA 632 process of solution definition, we will then examine the way in which the different qualitative and quantitative safety requirements can be derived and assigned to components and to the structure of the system under consideration to generate specified safety requirements.

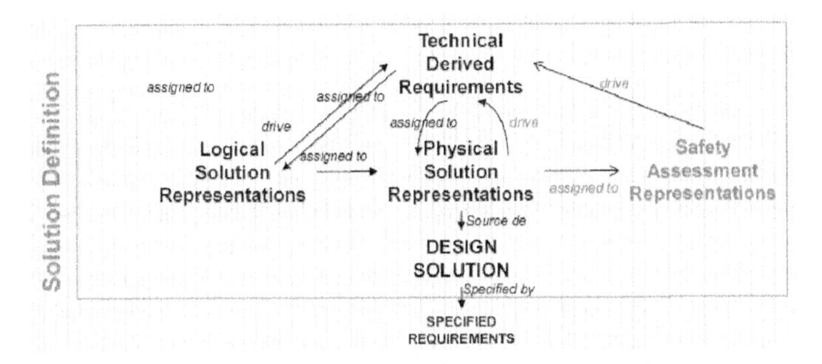

Figure 10.7. *EIA632 solution definition process extended to safety aspects. For a color version of this figure, see www.iste.co.uk/micouin/MBSE.zip*

We have introduced the derivation mechanism enabling the replacement, in the design process, of a system level PBR by the conjunction of derived requirements (PBRs) assigned to all or a part of its subsystems in section 7.2.

For such a derivation $Rq \rightarrow \{Rq_1,..,Rq_n\}$ to be true, the following formal relationship shall be established:

$$\text{when DC} \rightarrow \text{Rq} \leq \text{Rq}_1 \wedge .. \wedge \text{Rq}_n \quad [10.1]$$

This relationship [10.1] means that the conjunction of derived requirements Rq_i assigned to subsystems $s\Sigma_1, \ldots, s\Sigma_n$ of a system Σ shall be more constraining than the system requirement Rq it is derived from, if the design choices (DCs) are respected.

The derivation of a safety requirement *Rq* assigned to a type of systems Σ complies with this general mechanism and produces a set of derived safety requirements Rq_i assigned to all or a part of its subsystems ($s\Sigma_1, \ldots, s\Sigma_n$). This derivation essentially depends on the structure of Σ, that is to say its architecture. Any change in the structure usually makes the derivations depending on it obsolete.

For system safety considerations, we are often brought to consider typical structures (patterns) on which the analyses are based, because they enable assessments based on different probability calculi theorems or technological operative rules, and on which we can base ourselves to derive quantitative and qualitative safety requirements.

Table 10.5 shows some of the most famous design patterns; many other patterns can be considered such as the "stand-by redundant systems", "load sharing systems" and "non-parallel-series systems" patterns (see [VIL 88] or [MOD 10]).

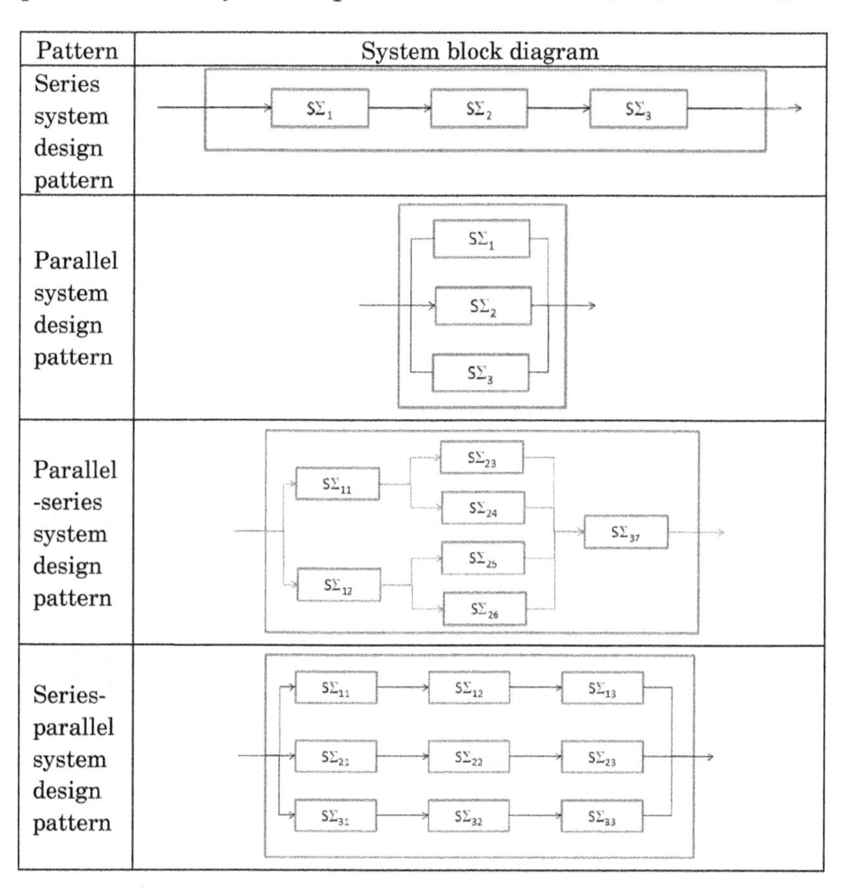

Pattern	System block diagram
Series system design pattern	$S\Sigma_1$, $S\Sigma_2$, $S\Sigma_3$
Parallel system design pattern	$S\Sigma_1$, $S\Sigma_2$, $S\Sigma_3$
Parallel -series system design pattern	$S\Sigma_{11}$, $S\Sigma_{12}$, $S\Sigma_{23}$, $S\Sigma_{24}$, $S\Sigma_{25}$, $S\Sigma_{26}$, $S\Sigma_{37}$
Series-parallel system design pattern	$S\Sigma_{11}$, $S\Sigma_{12}$, $S\Sigma_{13}$, $S\Sigma_{21}$, $S\Sigma_{22}$, $S\Sigma_{23}$, $S\Sigma_{31}$, $S\Sigma_{32}$, $S\Sigma_{33}$

Table 10.5. *Examples of system design patterns considered for safety aspects*

If we consider the following quantitative safety requirements:

Rq_{Σ}: $val(\Sigma.failure\_rate) \leq \Lambda_{\Sigma}$

$Rq_{s\Sigma i}$: $val(s\Sigma_i.\ failure\_rate) \leq \lambda_{s\Sigma i}$

Then, the derivations for the patterns above (Table 10.5) are true:

when SeriesSystem $\rightarrow \mathrm{Rq}_{\Sigma} \leq \mathrm{Rq}_{s\Sigma 1} \wedge \mathrm{Rq}_{s\Sigma 2} \wedge \mathrm{Rq}_{s\Sigma 3}$ *iff* $\Lambda_{\Sigma} \leq \lambda_{s\Sigma 1} + \lambda_{s\Sigma 2} + \lambda_{s\Sigma 3}$ (1)

when ParallelSystem $\rightarrow \mathrm{Rq}_{\Sigma} \leq \mathrm{Rq}_{s\Sigma 1} \wedge \mathrm{Rq}_{s\Sigma 2} \wedge \mathrm{Rq}_{s\Sigma 3}$ *iff* $\Lambda_{\Sigma} \leq \max(\lambda_{s\Sigma 1}, \lambda_{s\Sigma 2}, \lambda_{s\Sigma 3})$(2)

when Parallel-SeriesSystem $\rightarrow \mathrm{Rq}_{\Sigma} \leq \mathrm{Rq}_{s\Sigma 11} \wedge .. \wedge \mathrm{Rq}_{s\Sigma 7}$
iff $\Lambda_{\Sigma} \leq \max(\lambda_{s\Sigma 11}, \lambda_{s\Sigma 12}) + \max(\max(\lambda_{s\Sigma 11}, \lambda_{s\Sigma 12}), \max(\lambda_{s\Sigma 11}, \lambda_{s\Sigma 12})) + \lambda_{s\Sigma 7}$ (3)

when Series-ParallelSystem $\rightarrow \mathrm{Rq}_{\Sigma} \leq \mathrm{Rq}_{s\Sigma 11} \wedge .. \wedge \mathrm{Rq}_{s\Sigma 33}$
iff $\Lambda_{\Sigma} \leq \max(\lambda_{s\Sigma 11}+\lambda_{s\Sigma 12}+\lambda_{s\Sigma 13}, \lambda_{s\Sigma 21}+\lambda_{s\Sigma 22}+\lambda_{s\Sigma 23}, \lambda_{s\Sigma 31}+\lambda_{s\Sigma 32}+\lambda_{s\Sigma 33})$ (4)

In the same way, if we consider the following qualitative safety requirements:

Rq_{Σ}: $val(\Sigma.DAL)=X$ *with* $X \in \{A, B, C, D, E\}$
$Rq_{s\Sigma i}$: $val(s\Sigma_i.DAL_i) \leq x_i$ *with* $x_i \in \{A, B, C, D, E\}$

The following derivations, which are an interpretation of Table 3 of the ARP4754A, for the patterns *SeriesSystem* and *ParallelSystem* are true:

when SeriesSystem $\rightarrow \mathrm{Rq}_{\Sigma} \leq \mathrm{Rq}_{s\Sigma 1} \wedge \mathrm{Rq}_{s\Sigma 2} \wedge \mathrm{Rq}_{s\Sigma 3}$ *iff* $\forall i,\ x_i = X$ (1)

when ParallelSystem $\rightarrow \mathrm{Rq}_{\Sigma} \leq \mathrm{Rq}_{s\Sigma 1} \wedge \mathrm{Rq}_{s\Sigma 2} \wedge \mathrm{Rq}_{s\Sigma 3}$ *iff*

$\{\exists i\ x_i=X,\ \forall j \neq i,\ x_j=succ(succ(X))\}$ or $\{\exists i,j\ j \neq i,\ x_i= x_j=succ(X),\}$

10.3.4. *Simulation and validation of the derived safety requirements*

The process of validation by simulation of derived requirements described in section 8.3.4 is entirely applicable to the derivation of quantitative and qualitative safety requirements, such as we defined them above.

The goal of this validation is to ensure the correctness of these derivations and their completeness. It will ensure the satisfaction of the system level safety requirements for a given structure of the system, if the derived safety requirements are satisfied.

The levels of rigor mentioned in section 8.3.5 are applicable. We will note that the determination of the effect resulting from the components' structural properties on the corresponding system level structural properties, requires the introduction, within the structural design model, of virtual components, able to compute these effects.

We define these components as virtual, in the sense that they correspond to no real, tangible, objects, but are the models (approximately true) of the, no less real, laws of physics.

This is particularly the case for the assessment of failure rates per flight hour of a type of systems, which depends, on the one hand, on the failure rate per flight hour of each of its components and, on the other hand, of the system architecture, as illustrated for a system in sequence in Figure 10.8.

In this first structural design model, the virtual component "failure rate" encapsulates the statistical law linking the failure rate of a system in series to that of its physical components.

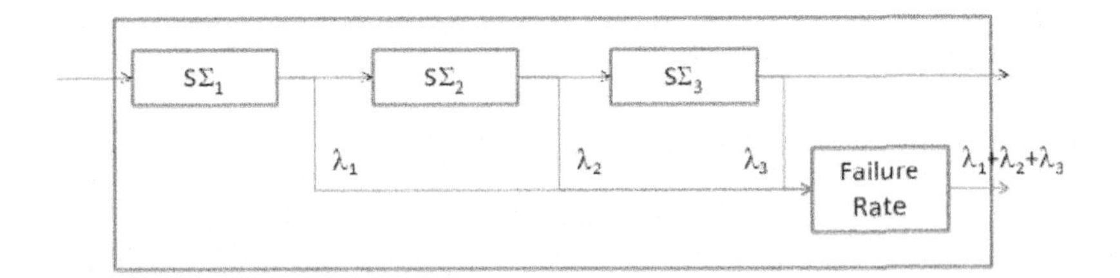

Figure 10.8. *SDM virtual component computing failure rates*

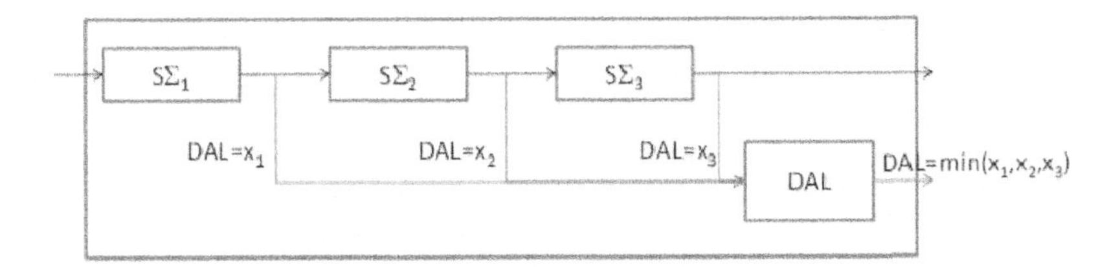

Figure 10.9. *SDM virtual component computing DAL*

This is also the case for the assessment of the FDAL of a type system, which depends, on the one hand, on the F/IDAL of each of its components and, on the other hand, on the system architecture, as illustrated for a system in sequence in Figure 10.9.

For this example, the structural design model embeds the virtual component "DAL", encapsulating the technological operative rule fixing the FDAL of a system in series as a function of the physical components F/IDAL.

As we can imagine, these virtual components can be used to determine the system level resultant of many structural properties, such as mass balance, inertia balance and electric consumption balance. Often, they are linked to the structure of the type of systems under consideration. A noteworthy exception concerns the mass balance, which is independent of the type of systems structure, and only depends on the constitutive components.

10.3.5. *Simulation and verification of the failure prevention mechanisms*

The verification process by simulation of the design models with regard to the specification model described in Chapter 9 (section 9.3) is entirely applicable to the verification of the design with regard to the quantitative and qualitative safety requirements, such as we described them above.

The goal of this verification is to ensure the safety properties (effective F/IDAL or effective failure rates) of the components such that they go back up the development and assessment "bottom-up" processes, satisfy the component level safety requirements and enable the integrated system to satisfy its own requirements (FDAL or failure rate).

They also enable us to verify by simulation, with the required level of rigor, that the mechanisms introduced at the system structural level to prevent or mitigate failures being efficient, that is to say the mechanisms enable the system to tolerate single or combined faults, while ensuring the required function without failure.

They finally enable us to verify by simulation, with the required level of rigor, that the reconfiguration mechanisms designed to passivate the faults and ensure the function to be carried out are maintained, and respond effectively.

There are several safety mechanisms being considered, and they notably include redundancies, segregations, dissimilarities, feedback loops, voting and reconfiguration mechanisms.

This kind of verification requires setting up the last type of design models manipulated in the Property Model Methodology (PMM), which we present in the following section, namely the reliability design models (RDMs) inspired from the reliability block diagrams (see [VIL 88] or

[MOD 10]) used, instead of (or complementarily to) the failure trees method.

10.3.6. *Reliability design models*

RDMs are models designed to verify the robustness of a system regarding internal faults.

We can, according to the needs, consider single fault or even combined faults to measure their effects (failures), verify the intended effectiveness of the fault passivation and reconfiguration mechanisms. They can also enable the detection of common fault modes linked to the considered architecture of the type of systems under consideration.

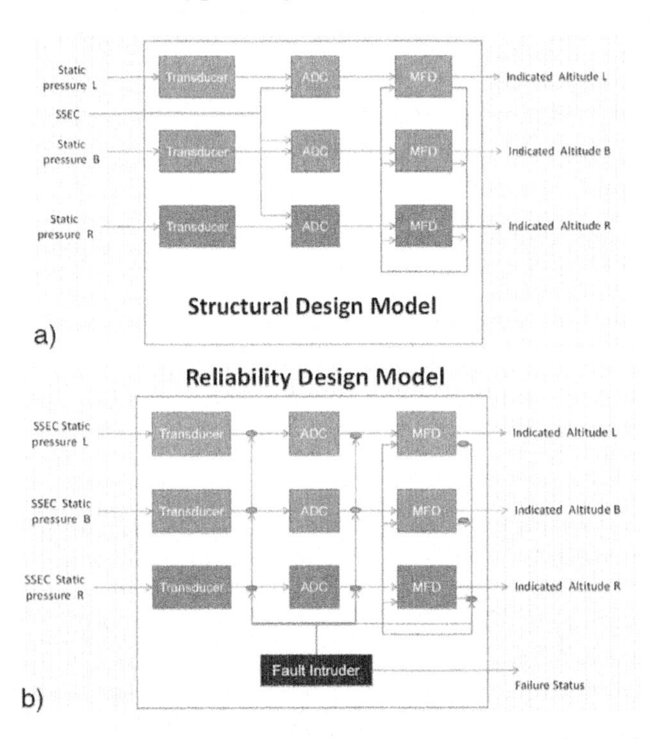

Figure 10.10. *Reliability design model b) derived from a baro-altimeter structural design model a). For a color version of this figure, see www.iste.co.uk/micouin/MBSE.zip*

An RDM model, in Figure 10.4(b), derives directly from a structural design model (a). The analysis of this structural design model enables the systematic identification of internal faults of the type of system under consideration. These faults manifest at each component output. Thus, in Figure 10.4(a) a "transducer", an "ADC" or an "MFD" can lose its output flow, or even produce an undetected faulty output flow. This example highlights 18 single faults.

An additional virtual component, the "Fault Intruder" enables the intrusion of simple or combined faults by corrupting the outputs, according to the needs and objectives assigned to the RDM. Thus, in the example, during the execution of a simulation scenario designed to verify the tolerance of an architecture to single faults, the "Fault Intruder" can inject one of the 18 single faults identified, which will successively corrupt the different outputs from different components constituting the system. In this way, we can ensure the ability of the system structure to tolerate single faults.

Although no general common cause fault detection mechanism exists, the RDMs can enable, on the one hand, the detection of those linked to the structure while, on the other hand, model configuration mechanisms can cover dissimilarity requirements between components through an assessment of the configuration indexes.

10.3.7. *Safety theorem: validating additional requirements*

We conclude this chapter with a specific point, related to the introduction of additional requirements at the subsystem specification model level, and its impact on the safety of the type of systems that includes such a subsystem.

A rule of the art dictates that the global impact on the safety of the encompassing system shall be evaluated when some additional requirements are introduced at the subsystem level. This assessment is necessary when working with textual specifications, particularly when the system specification has weaknesses, which are compensated by additional subsystem requirements.

However, we will show that this assessment becomes superfluous so long as the PMM and PBRs are used.

This is the safety theorem: the introduction of any additional PBR at the component level of a type of systems has no effect on the safety of that type of systems.

To demonstrate this theorem, we consider a feasible type (that is to say non-empty) Σ of systems, specified by a set of PBRs $\{PBR_1, .., PBR_n\}$. We have $\Sigma = SAT(\{PBR_1, .., PBR_n\}) \neq \emptyset$. We now suppose that these systems present a failure F. We characterize this failure in the following way: when a condition C is satisfied, a property P is in a domain D, considered as unacceptable (for example, because its consequences would be catastrophic), in other words, the faulty state is constituted of the conjunction of the condition C and the fact P is in D ($F = C \wedge (P \in D)$). To prevent this failure, we introduce the safety requirement:

PBR_{safety} : when $C \Rightarrow P \notin D$

It then follows that the systems of type Σ, which satisfy this PBR_{safety} requirement, define a subset Σ' of Σ such that $\Sigma' = SAT(\{PBR_1, .., PBR_n, PBR_{safety}\})$.

If the systems of the subset Σ' are feasible ($\Sigma' \neq \emptyset$), then they are protected against the failure F.

We now suppose that Σ' is composed of (DC) the subsystems $O_1, .., O_p$. Then, each PBR_i of $\{PBR_1, .., PBR_n,$

$PBR_{safety}\}$ applied to Σ' has been derived into PBR_{ij} or $PBR_{safetyj}$ applied to all or a part of the components O_1, .., O_p, according to DC.

If all the derivations of PBR requirements are true (valid), then the conjunction of requirements assigned to Σ' is less constraining than the conjunction of all the requirements assigned to the different components of Σ', that is to say:

when DC => $PBR_1 \wedge .. \wedge PBR_i \wedge .. \wedge PBR_n \wedge PBR_{safety}$

$\leq (PBR_{11} \wedge .. \wedge PBR_{1j} \wedge .. \wedge PBR_{1p} \wedge PBR_{safety1})$

$(PBR_{i1} \wedge .. \wedge PBR_{ij} \wedge .. \wedge PBR_{ip} \wedge PBR_{safetyj})$

..

$(PBR_{n1} \wedge .. \wedge PBR_{nj} \wedge .. \wedge PBR_{np} \wedge PBR_{safetyp})$

The result of this is therefore that a system, with an architecture compliant to DC and where each component O_i is compliant with the requirements $\{PBR_{i1}, PBR_{ij}, PBR_{ip}, PBR_{safetyj}\}$, is protected against failure F.

If, finally, a PBR_{add} requirement is added to the specification of component O_i, for any reason, we will still have:

$(PBR_{i1} \wedge .. \wedge PBR_{ij} \wedge .. \wedge PBR_{ip} \wedge PBR_{safetyj}) \leq PBR_{i1} \wedge .. \wedge PBR_{ij} \wedge .. \wedge PBR_{ip} \wedge PBR_{safetyj}) \wedge PBR_{add}$

and the relationship remains true.

when DC => $PBR_1 \wedge .. \wedge PBR_i \wedge .. \wedge PBR_n \wedge PBR_{safety}$

$\leq (PBR_{11} \wedge .. \wedge PBR_{1j} \wedge .. \wedge PBR_{1p} \wedge PBR_{safety1})$

$(PBR_{i1} \wedge .. \wedge PBR_{ij} \wedge .. \wedge PBR_{ip} \wedge PBR_{safetyj}) \wedge PBR_{add}$

..

$(PBR_{n1} \wedge .. \wedge PBR_{nj} \wedge .. \wedge PBR_{np} \wedge PBR_{safetyp})$

In other words, a system with an architecture compliant with DC and where each component O_i is compliant with the requirements $\{PBR_{i1}, PBR_{ij}, PBR_{ip}, PBR_{safetyj}\}$ and to which we have added additional requirements remains protected against failure F.

NOTE.– If the introduction of any additional PBR at the component level of a type of systems has no effect on the safety of that type of systems, it remains true that this introduction has an effect on its feasibility. Indeed, the introduction of a new PBR always has an effect of reducing:

1) the set of solutions satisfying the specification;

2) the state space of the type of systems under consideration.

which converge to the empty set ∅, that is to say toward the non-feasibility, if we do not pay attention to it.

10.4. Conclusion

As we have seen it previously, the PMM is a strongly integrated approach of development processes (specification and design), requirements validation, design verification and safety assessment processes.

This brings us to propose an extension of the EIA 632 design process by adding to it, and at different levels such as the requirement definition process and solution definition process, activities related to safety assessment.

In Figure 10.11, on the one hand, a safety requirement determination activity is integrated in the EIA 632 requirement definition process, and on the other hand, an activity defining the safety assessment representation, the RDMs from which it is possible to carry out safety requirement validation activities and activities verifying the design model in relation to the safety requirements, is integrated in the EIA 632 solution definition process.

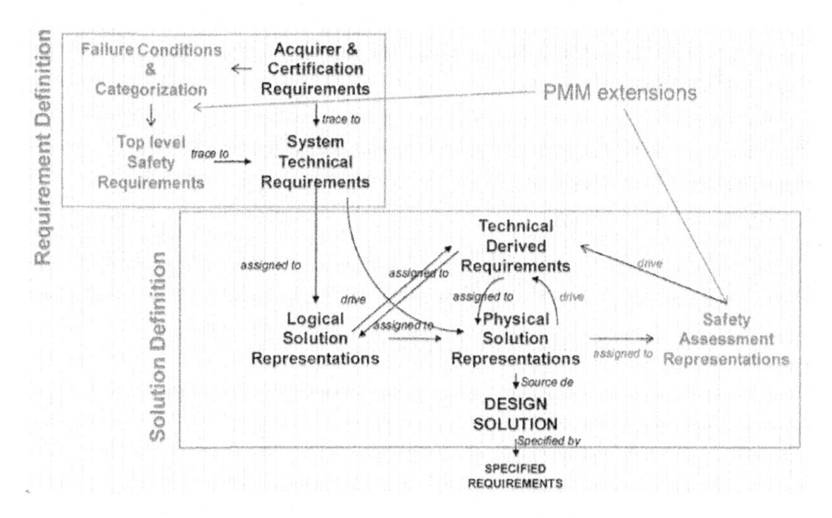

Figure 10.11. *Extended EIA 632 design process model. For a color version of this figure, see www.iste.co.uk/micouin/MBSE.zip*

11

Property Model Methodology Development Process

11.1. Introduction

In this chapter, we will describe the system development process promoted by the property-model methodology (PMM). In order to introduce this process, we will first describe its upstream environment. And then, we will present the different steps of such a development. Then, we will describe each step in detail and, to conclude, we will discuss issues related to configuration management and collaborative work within the framework of the PMM.

11.2. Early phase of a system development, preliminary studies

Developing a new type of systems is usually preceded by a preliminary studies phase. During this preliminary studies phase, we first consider the practical and economical interests of these type of systems. Second, the feasibility and efficiency of such a type of systems is also assessed. The aim is not, at this step, to work through a systems industrial development process, but to highlight the operational environment conditions, operational goals, operating principle of the new type of systems and an associated operational concept. However, if the type of systems under consideration is derived from precedent type of systems, which already exists, this preliminary study phase can be reduced to a re-evaluation of the parameters taken into account during the initial preliminary studies. During these studies, it is not rare to set up simulation methods and develop models. This preliminary studies phase is not covered in the framework of PMM, but it would not be without interest that models drafted during the phase be industrialized in the framework of the system development.

If, for example, these preliminary studies relate to a midair collision avoidance system (CAS), installed in an aircraft, these studies will first be focused on the conditions of midair traffic, namely the traffic density of the different air-space zones. This is done to appreciate the risks of collisions, today and tomorrow, in relation to the foreseeable evolution of traffic, the inacceptable, or not, character of these risks, and to define the safety objectives capable of containing these risks within socially acceptable limits. The preliminary studies will have to characterize the goal of the type of systems, namely, enable the avoidance of aircraft collisions in the air, with a success rate compatible with the safety objectives. Preliminary studies will then consider operational strategies (detection, place of operators in the loop, warning operators, coordination between aircraft involved and characterization of avoidance maneuvers) by plugging them into operation scenarios. The efficiency of these strategies will have to be evaluated with regard to (1) the safety objectives and (2) the available midair collision data or risky situation data, while the situations exhibiting flaws in these strategies will be identified. The preliminary studies will have to identify the resources that are necessary and available to carry out these strategies, as well as the impact of deploying such a new type of systems on the operators, in operational and organizational terms.

Preliminary studies can be concluded with an operational[1] and maintenance concept, which the industrial development of the type of systems under consideration can be based on.

11.3. Steps of the industrial development of a type of systems

The concepts of operation and maintenance provide a possible entry point to the industrial development of the type of systems under consideration. The PMM covers this industrial development, from a methodological perspective, by decomposing it into different steps as illustrated in Figure 11.1.

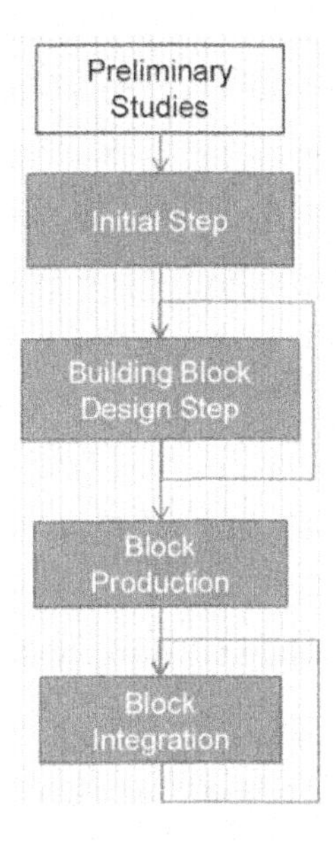

Figure 11.1. *PMM system development process*

1 IEEE Std 1362-1998, IEEE Guide for Information Technology – System Definition – Concept of Operations (ConOps) Document.

The following steps are prescribed by the PMM:

1) initial step: highest level system specification;

2) design step: descending and iterative design of building blocks down to the lowest level;

3) production step: lowest level building block production;

4) integration step: ascending and iterative integration of building blocks up to the integrated system and installed in its operational environment.

In the following sections, we will describe these different steps by identifying the different activities and fail/pass criteria for each step.

Figure 11.2. *Colossus with feet of clay in Nebuchadnezzar's dream*

11.4. Initial step: highest level system specification

The highest level system specification is the cornerstone of a descending development process of a type of systems. It provides the reference for the whole of the development. Consequently, it also concentrates the maximum risk of transforming development into a "colossus with feet of clay[2]". If the system specification is not carried out well, blurred, incorrect or incomplete, we can fear that the resulting systems will be unsuitable to the needs, that their development costs will strongly surpass the objective costs and that the provision deadline will not be respected.

11.4.1. *Initial step general approach*

The proposed approach for this first step aims to reduce the risks, although it cannot be completely eliminated. It is described according to the following six points:

1) First, a specification model of the type of systems under consideration is established. Most often, it is recommended to develop this specification model incrementally, in successive versions, in order to correctly identify the different capabilities of the type of systems, one after the other.

2) For each increment of the specification model, a preliminary analysis of the functional hazards (FHAs) can be carried out (see section 10.3.2), which helps clear out the failures associated with the different capabilities taken into account in the model (loss of the capability and undetected degraded capability), and helps categorize their severity according to the foreseeable consequences of the failure.

3) This categorization of the severity of the identified failures helps determine (i) the level of rigor with which the validation of the specification model shall be carried out for the modeled capabilities, (ii) the scenarios and validation conditions that must be run through to ensure the specification exactness of the capabilities taken into account

2 "32-Then was the iron, the clay, the brass, the silver, and the gold, broken to pieces together, and became like the chaff of the summer threshingfloors; and the wind carried them away, that no place was found for them" Daniel, Chapter 2–32.

and (iii) enables us to add safety requirements in order to prevent the identified risks. The scenarios and validation conditions are devised in accordance with the level of rigor required (section 8.3.5).

4) The equation design model (EDM) (see section 7.4.2) shall be derived from the system specification model (SSM) to be validated and a validation workbench (see section 8.3.3) composed of a system model (SM) that includes the specification model (SSM) to be validated and its EDM and a validation driver.

5) The SM shall be simulated in the validation workbench, on the basis of the scenarios and validation cases defined above, which are submitted to the SM by the validation driver. This simulation enables an authorized person (a pilot or a test engineer, an expert from the authority, a client, etc..) (1) to observe the behavior of the type of specified systems, and (2), if it is the case, to confirm that the observed behavior is consistent with the intended behavior (for each scenario and validation condition considered) or (3) if it is not the case, declare the observed behavior as invalidated.

6) If some observed behavior has been invalidated, the specification model shall be revised in order to correct the observed discrepancies and return over the process indicated above. When no more discrepancies are observed, we can declare that the specification model, in the version under consideration, is the most possibly exact model for the pre-established level of rigor.

11.4.2. *Establishing a specification model of the type of systems*

Specifying a type of systems Σ according to the PMM approach involves constructing specification models. It is a gradual development process of the system model that can require multiple iterations based on the understanding that it is better to quickly dispose of a crude ("false") model to then rework it, rather than suffer the absence of a model. Simulation can, therefore, be a means to adjust the specification model, which will progressively mature through successive improvements.

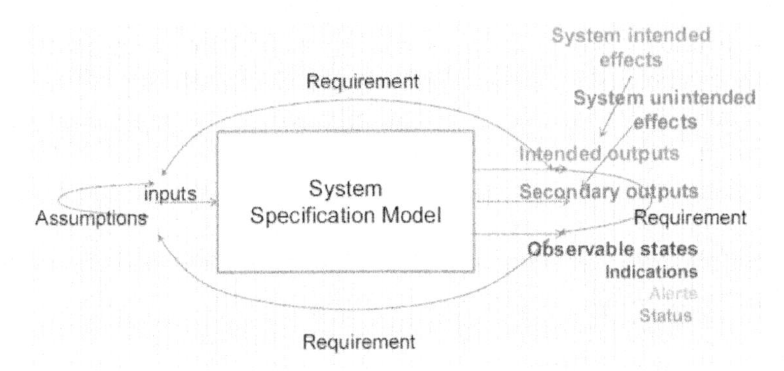

Figure 11.3. *System specification model. For a color version of this figure, see www.iste.co.uk/micouin/MBSE.zip*

According to PMM, a type of systems is specified by its outputs. It is useless to progress any further as long as we do not know what output the system has to produce.

Thus, the first thing to do is to focus on the effects it shall produce on its environment, that is to say its functions (see section 2.3, Function=Intended Effects of a system). We can determine them by, for example, considering the results of preliminary studies and the operational concept that comes from it. We must temporarily forget the inputs, we will return later. Focusing on outputs, we must ask questions such as: (1) what kind shall these outputs be (matter, power, data or events flows)? (2) do these flows have to be continuous or discrete? (3) what are the rates at which these outputs shall be produced? (4) if so, what rate of change in these outputs shall be ensured? We can at the same time, or later, be brought to specify the failure conditions (see section 10.3.2) such as the loss of one or more outputs and detected error outputs (what of the undetected error outputs?).

For example, if we specify a collision avoidance type of systems, installed in an aircraft to prevent midair collisions, the first thing to do is to determine the intended outputs of the systems: these outputs can be (1) visually indicate the crew of the aircraft equipped with a CAS, all aircraft within its immediate vicinity, (2) visually warn the crew and emit an aural warning when an intruder aircraft is closer, finally (3) suggest the crew an avoidance maneuver if the risk of collision is found. In addition to the functional outputs specification, we can imagine the specification of a CAS observable state, including, for example, failure codes. Then, the question of knowing how many intrusions the system shall be able to take into account concurrently comes? This being set, we know what the system shall do.

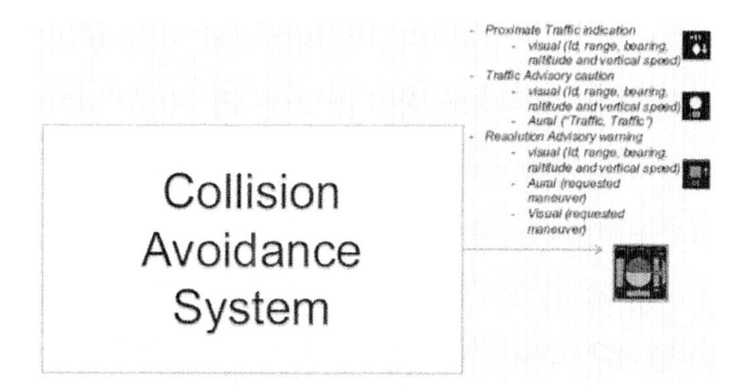

Figure 11.4. *CAS specification model of intended functions. For a color version of this figure, see www.iste.co.uk/micouin/MBSE.zip*

The outputs being defined, we can now focus on the conditions the outputs shall be performed in. There are multiple kinds of conditions. These could be, for example, an output must appear within a given timeframe after the occurrence of an input, and this output must be a function of the input. Another possibility is that an output must appear within a given timeframe after the occurrence of an observable state, and this output is a function of the observable state under consideration. We can multiply, as needed, the appearance conditions of an output and its current value. That is to define the inputs or the observable states linked to the outputs by property-based requirements.

For example, if we specify a type of CASs installed in aircraft to prevent midair collisions, the outputs having already been identified, we shall specify the appearance and disappearance conditions of a (1) close traffic indication, of a (2) traffic warning and, of (3) a caution associated with a conflict resolution advice, and all those conditions depend on a representation of the aircraft local environment, elaborated by the system. This local environment representation constitutes an observable state of the model that conditions its outputs and shall be as exact as possible.

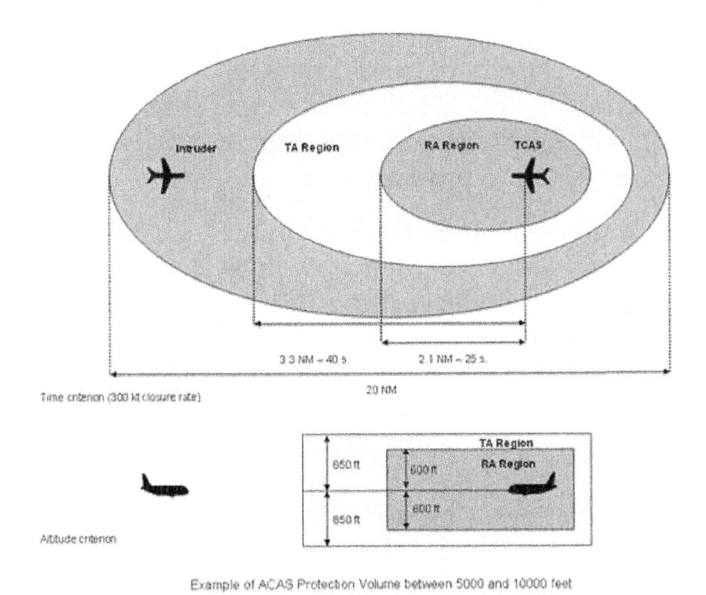

Figure 11.5. *CAS protected volume and local environment representation. For a color version of this figure, see www.iste.co.uk/micouin/MBSE.zip*

After this, we are brought to ask ourselves the same questions about the observable states as the ones asked about the outputs, that is to say, what their appearance and elaboration conditions are. For example, if we specify a type of CASs installed in aircraft to prevent midair collisions, the outputs having already been established, we will specify the elaboration and conditions of the local environment representation the system builds. This representation of the local environment is elaborated from inputs, from the CAS carrier aircraft and from aircraft located in the CAS carrier aircraft close environment.

Once the outputs and observable states have been taken care of, the case of inputs remains. What is the nature of the inputs, are they material flows, energy flows or even information flows? Are these flows continuous or discrete? What are the magnitudes or definition domains? What are the flow rates or the highest frequencies of these inputs? What are the logics of appearance or disappearance of these inputs? If relevant, what are the variation rates of these inputs? As these inputs come from the environment, we do not have control over them. In this case, we do not formulate requirements but assumptions (that are formally identical but very different in practice).

For example, for a CAS, these inputs can be a number of physical aircrafts with their own kinematics (position and speed). We will, therefore, formulate assumptions on (1) the maximum number of aircrafts that can be taken into account by the CAS, (2) the accessible information concerning these aircraft. Thus, we obtain assumptions on the available inputs of the type of systems and conversely the operational limits of this type of systems. We will also be able to imagine a CAS, based on the data provided by an automatic-dependent surveillance-broadcast (ADS-B)[3] system, as an input assumption.

---

3 ADS-B = automatic-dependent surveillance-broadcast.

Once the outputs, observable states and the inputs of the type of systems to be developed are identified, the relationships, which shall exist, between them shall be specified. In other words, how the outputs shall be devised as functions of the observable states and/or inputs. In which circumstances shall an event appear as an output, as a function of which input events? How shall the value of an output be devised from the value of the inputs and/or observable state? What shall the temporal inertia of the type of specified system be? What maximum delay can it introduce between the coming of the inputs and the appearance of the outputs?

For example, for a CAS, the specification of the relationship between the inputs and outputs can be the following:

1) The position of a new intruder shall be visually reported to the crew each time an aircraft enters the green envelope defined around an aircraft equipped with a CAS (own aircraft) in Figure 11.5.

2) The position of the intruders shall be visually presented to the crew so long as they remain within the green envelope in Figure 11.5.

3) The position of an intruder shall cease to be visually presented to the crew when it leaves the green envelope in Figure 11.5.

4) A traffic advisory (TA) alert shall be visually and aurally emitted to the crew each time an aircraft enters the yellow envelope defined around an aircraft equipped with a CAS in Figure 11.5.

5) A resolution advisory (RA) alert, including an appropriate maneuver message, shall be visually and aurally emitted to the crew each time an aircraft enters the red envelope defined around the aircraft equipped with a CAS in Figure 11.5.

6) An aural message “clear of conflict” shall be emitted to the crew when the threat leaves the red envelope in Figure 11.5.

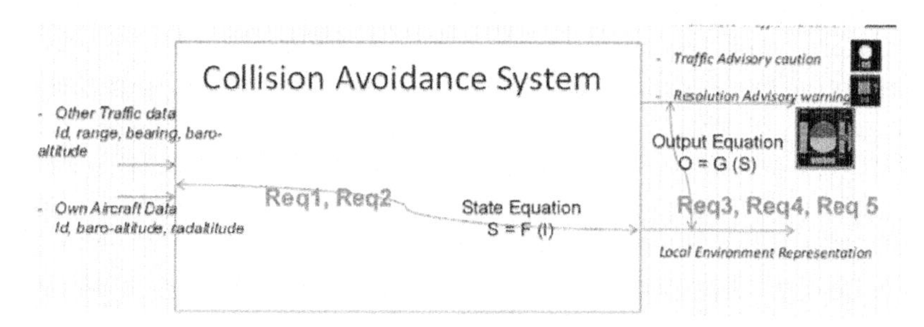

Figure 11.6. *CAS specification model PBRs. For a color version of this figure, see www.iste.co.uk/micouin/MBSE.zip*

This gives us six property-based requirements:

PBR_a: when CAS_Engaged and inGreen_Envelop(NewIntruder)

→AdditionalIntruderDataDisplayed

PBR_b: when CAS_Engaged and inGreen_Envelop(Intruders)

→IntruderPositionDisplayUpdate

PBR_c: when CAS_Engaged and goesoutGreen_Envelop(Intruder)

→IntruderPositionDisplayCanceled

PBR_d: when CAS_Engaged and inYellow_Envelop(Intruder)

→TA_Displayed and Aural(“Traffic, Traffic”)

PBR_e: when CAS_Engaged and inRed_Envelop(NewIntruder)

→RA_Displayed and Aural(Avoidance_Maneuver)

PBR_f: when CAS_Engaged and goesoutRed_Envelop(Intruder)

→Aural(“Clear of conflict”)

The terms appearing in this specification model all refer to a state (a value of a property) of the targeted type of systems. This is what we named the semantic assumptions (see section 4.3) of a model.

SA_1: *CAS_Engaged: the CAS system is installed and engaged*

SA_{21}: *inGreen_Envelop(NewIntruder): a new aircraft has been detected in the external (green) envelope*

SA_{22}: *AdditionalIntruderDataDisplayed: the characteristics of the last aircraft detected in the external (green) envelope are visually presented to the crew*

SA_{23}: *inGreen_Envelop(Intruders): the characteristics of the aircrafts detected in the external (green) envelope evolve*

SA_{24}: *IntruderPositionDisplayUpdate: the visual presentation, intended to the crew, of the characteristics of the aircraft detected in the external (green) envelope is updated*

SA_{25}: *goesoutGreen_Envelop(Intruder): an aircraft leaves the external (green) envelope*

SA_{26}: *IntruderPositionDisplayCanceled: the visual presentation, intended to the crew, of the characteristics of the aircraft detected in the external (green) envelope is deleted*

SA_{31}: *inYellow_Envelop(NewIntruder): a new aircraft has been detected in the intermediary (yellow) envelope*

SA_{32}: *TA_Displayed: a visual Traffic Advisory alert is emitted for the crew*

SA_{33}: *Aural("Traffic, Traffic"): the aural alert "Traffic, Traffic" is emitted for the crew*

SA_{41}: *inRed_Envelop(NewIntruder): a new aircraft has been detected in the internal (red) envelope*

SA_{42}: *RA_Displayed: a visual Resolution Advisory alert is emitted for the crew*

SA_{43}: *Aural(Avoidance_Maneuver) : the aural alert indicating the type of maneuver to be carried out is emitted for the crew*

SA_{51}: *goesoutRed_Envelop(Intruder): an aircraft leaves the internal (red) envelope*

SA_{52}: *Aural("Clear of conflict"): the aural alert indicating the end of the conflict is emitted to the crew*

Of course, this specification is objective but it is not necessarly exact. It shall be validated.

Finally, we know that a system does not only produce intended effects but it can also produce undesired effects, which we will seek to limit, false alarms on one hand, but also other unwanted events on the other hand (see section 10.3), in order to make them tolerable or acceptable.

11.5. Design steps: descending and iterative design of the building blocks down to the lowest level blocks

Once we have a specification model of the type of systems under consideration that is sufficiently exact (in other words, validated as indicated above), it is possible to start the design phase, which is decomposed into different design steps by obeying the following rules:

1) Each design step corresponds to a building block: for example, a system, subsystem, item, component, assembly and part of the type of systems considered.

2) Dependence rule: the design step of a block B can only be engaged after the completion of its specification model (see section 6.7).

3) Independence rule: the design steps of two blocks B_1 and B_2 can be engaged independently from one another when B_2 is not in B_1 lineage, and conversely.

4) Two variations of the design process cover, on the one hand, non-terminal blocks (for example, a system, subsystem, item, component or assembly) and, on the other hand, the lowest level blocks (terminal) or parts. The first kind is described in the following section. The second is described right after.

11.5.1. *Design step of a non-terminal block*

In case we have a clear vision of the behavioral chain to be set up in order to satisfy the requirements allocated to a non-terminal block, a design step of this block can, straightaway, be limited to a structural design, so long as we dispose of a validated specification model for that block.

1) In this case, we define a structural design model (SDM) of this block (see section 7.6.2) as follows. Usually, we develop such an SDM by taking into account the general architectural constraints linked to (1) the knowledge we have of the design patterns established by the state of the art, (2) the components and links existing on the market and (3) the installation and environmental constraints.

i) Selecting the components: the block designers are brought to consider different parts, assemblies, components, items or subsystems, which could form the composition of the block under consideration. This choice is made depending on the knowledge the designers have of the domain, existing solutions, their advantages and disadvantages and the innovations that could be introduced.

ii) Defining the block exo-structure: some of the block components shall be linked to the inputs, observable states and outputs of the block, by material, energy or data links, in order to connect them to the block exo-structure.

iii) Defining the block endo-structure: in a complementary way, the components of the block shall be linked to one another, by material, energy or data links, and according to adapted architectural patterns, in order to constitute the block endo-structure.

2) Defining the specification models of the block components:

i) A block SDM having been established, the block designers derive the specification model requirements over the different selected components of the structural architecture of the block. Each specified requirement to the block is derived over on or more block components. Note that some block requirements, such as redundancy requirements, can be fully satisfied by the selected architectural pattern. They are not assigned to block components and are no more derived.

ii) The set of requirements derived and assigned to a block component form the specification model of that component. Additional requirements can be added to the specification model of this component with the potential effects clarified in section 10.3.7. The structural design solution is, therefore, constituted of a structural architecture, a set of specification models associated with the different components of the block under consideration.

3) Validating the specification model of the components: a final activity linked to the design of a block remains to be carried out, namely demonstrating that the requirements assigned to different components of the block have been correctly and completely derived from the requirements assigned to the block, and that the additional requirements added to some components do not cause infeasibilities. To do this, we only need to carry out a validation activity as we described in section 8.3.4. If this validation process is conclusive, we can then declare that the derivation and specification models assigned to the different components are valid. In the opposite case, the derivation shall be reworked to correct the identified errors.

If, on the contrary, the assumption we made at the start of this section, namely that the designer(s) have a clear vision of the behavioral chains to be set up in order to satisfy the requirements allocated to this block, it is more cautious to start the design step of this block with a behavioral design, from the moment we have a validated specification model of this block.

4) To do this, we only need to follow the rules described in the following section (11.5.2) to design a behavioral design model (BDM) of the block under consideration. Then, we need to ensure that the design model does not comprise any design errors in relation to the corresponding specification model according to the verification rules described in section 9.3.3.

5) Once the BDM is free of errors, we then have sufficient knowledge of the behavioral chains that could be set up to satisfy the requirements of the specification model of the block under consideration. We can then start the structural design of the block as indicated above by allocating one or more behavioral components (discrete, continuous or mixed processes) to each structural component (subsystem, item, component, assembly or part). This allocation of the behavioral design process shall be complete, i.e. each behavioral entity shall have a structural counterpart resulting of this allocation operation. Note that this allocation operation is not an injective operation, several behavioral entities may have a common structural counterpart. Conversely, the same behavioral entity may not have several structural counterparts in one design solution.

Remind that there generally is not a unique behavioral or structural solution corresponding to a block specification model. There can be none (unfeasible specification), but there are generally multiple, which can all be treated in the way indicated above.

11.5.2. *Behavioral design step of a terminal block*

We consider a block to be terminal (a block of the lowest level) within a system decomposition when it (1) is available off the shelf of a supplier (this is the case for, for example, with sold equipment with an approbation letter (TSO/ETSO)),

(2) is developed to satisfy defined specifications by a supplier and (3) can be directly made without any additional design. We recall also, that an equation-based model design is a terminal behavioral design model.

A behavioral design step of a lowest level block can start from the moment we dispose of a validated specification model of that block.

1) In this case, we define a BDM of that block as follows. Usually, we develop such a BDM by taking into account general architectural constraints, linked to the knowledge we have of design patterns and possible material transformations recognized by the state of the art. The preliminary studies can also have studied and developed transformation principles and algorithms validated by computation or simulations.

i) Selecting processes: the block designers are brought to consider different kinds of discrete, continuous or mixed processes, which could form the behavioral composition of the block under consideration. This choice is made according to the knowledge the designers have of the domain, the existing solutions, their advantages and disadvantages and the innovations that could be introduced.

ii) Defining the block exo-structure: some of the block processes shall be linked with the inputs, observable states and outputs of the block, by material, energy or data flows, in order to connect them to the block exo-structure.

iii) Defining the block endo-structure: in a complementary way, the block processes shall be linked to each other, by material, energy or data flows, in order to constitute the block endo-structure. This endo-structure constitutes the behavioral architecture of the block.

2) Verifying the block behavioral design model: a last activity linked to the behavioral design of a block remains to

be carried out, that is to say demonstrating that this behavioral design is feasible and without errors.

i) For this reason, using an EDM helps establish the specification model does not contain any absolute contradiction (see section 6.2). This is enough in the case of a terminal block.

ii) Using a BDM that is more complex than an EDM enables us to define functional chains while we design a non-terminal block, and that we seek to have a clear vision of these chains, as we mentioned in the previous paragraph. After identifying, as we indicate it in the previous indent, the design of an intermediary block can be taken up again as in section 11.5.1, indent 5.

11.5.3. *End of the design step*

The design phase ends with the verification of various level of integration of the system model ascending order design templates terminal blocks to model the complete system. This activity is fully described in section 9.3.3.

11.6. Realization step of the lowest level building blocks

The realization of a lowest level block can be subdivided into two substeps:

1) The block physical realization: it is not technically covered by the PMM. The manufacture of each lowest level block requires processes that are specific to the kind of block to be produced, machining a metal part, performing a special process for a composite part and so forth. However, processes in line with the PMM can be used when producing a software block or a complex electronic hardware (CEH). In this case, the PMM remains applicable, and a code generator or CEH synthesizer can be used, if they are available.

2) The verification of the realized block: once produced, the physical block shall be verified. In order to do this, the compliance with the specification model will be established on a testbench by confronting this block with the set of scenarios and verification cases that have been selected (see section 9.3.3). If this compliance is proven faulty, we can either suspect an error in the manufacturing process or in the material selection, either an unfeasibility of the specification model. In the first case, we can just correct the error, in the second; we have to modify the specification model to make it feasible. This modification of the specification model can impact the specification model of the blocks coupled with the preceding block within the encompassing model endo-structure. It can also impact the specification model of the encompassing block, and therefore, by domino effect, more or less severely impact the design of the type of systems under consideration. Wise designers limit this risk nonetheless by introducing components for which enough experimental feedback exists, or for which feasibility studies have been carried out, for example during the preliminary studies.

11.7. Integration and installation steps

The integration of a type of systems is an iterative and ascending process ending on an installation final step, which is composed of two substeps:

1) The physical integration of each non-terminal block through the assembly of its constituents, compliant with the endo-structure of the block under consideration. The integration of each non-terminal block requires processes specific to the kind of block to be integrated. However, processes in line with the PMM can be used when integrating a software block or CEH. In this case, PMM remains applicable, and a code generator or a CEH synthesizer can be used.

2) The verification of the block integration: once physically integrated, the block shall be verified. In order to do this, the compliance with the specification model will be established by confronting, on a laboratory or installed in an aircraft, this physical block with the set of scenarios and verification cases selected. If this compliance is proven faulty, we can suspect an error in the physical integration process. In this case, we shall correct the integration error or the installation error. It is a result we had anticipated within the framework of the contract theorem stated in section 9.3.5.

11.8. Conclusion

The development process we have described in this chapter allows mastering and sharing of the design and development workload, for a type of systems, so long as it clearly identifies the steps from which the workload can be shared between multiple separated teams, and performed in parallel. These sharing steps are specified in the introduction of section 11.5, indent 2 and 3, which provide the independence and dependence rules allowing the parallel set up of the workload, while the contract theorem, stated in section 9.3.5 ensures the correct integration of workloads carried out in parallel. Additionally, the development process proposed gives us the possibility to manage, in parallel, multiple different versions of the specification models and design models of different constituents of a building block, and therefore to configure the type of systems under consideration according to different clauses, which all ensure the same system-level requirements.

Appendix

A1.1. Introduction

The purpose of this Appendix is to propose elements to translate system models (SMs) following the property-model method (PMM) methodology into a simulation model that allows us (1) to validate the specification models, (2) to verify the PMM design models and (3) to prepare the underlying phases of verifying the realization (single physical verification, physical integration and installation verifications).

In section A1.2, we will discuss roles introduced by the PMM methodology, namely the roles of the specifier, designer, validator and verifier. We will also mention the means required to perform PMM activities, namely the front-end of modeling used by specifiers and designers and the back-end of simulation, for validators and verifiers.

In the next sections, we will cover the modeling of a system specification and modeling of different design models. In section A1.3, we will consider the declaration and definition of specification model ports as well as the declaration and definition of property-based requirements (PBRs), which are also included in a specification model. In section A1.4, we will consider the declaration and definition of different design models associated with a given specification model, namely equation design models, behavioral design models (BDMs), structural design and reliability design models (RDMs). For each of these models, the required resources are identified. The Appendix concludes with section A1.5, on the configuration of an SM using different building blocks (root, intermediate and terminal) that can compose it; each of them is an assembly of a specification and design model. The transformation of an SM into a simulation model allows us to, depending on the different configurations considered, guarantee the validation of a specification model, the verification of a design model and also to assess the efficiency of failure prevention or mitigation mechanisms.

A1.2. Roles and means

A1.2.1. *Roles*

The PMM methodology identifies, *a priori*, four main roles in the development process of a type of systems. This includes the following conceptual roles:

– specifier, whose role is to produce the specification model of a type of systems;

– designer, whose role is to produce the different necessary design models corresponding to a given specification model;

– validator, whose role is to define the validation scenarios and validation cases of a given specification model, and to validate this specification model;

– verifier, whose role is to verify the different stages (design, production, integration and installation) of the implementation of a type of systems that are error-free.

These four conceptual roles may be played by the same physical person or, on the contrary, by different physical persons of the same or different organizations depending on the availabilities and the level of independence defined in the validation and verification plans.

Additional roles may be considered, for example, a workbench administrator whose role is to continuously maintain a description of modeling and simulation tools, the recording of the problems and the evolutions for qualified tool (or not) to ensure the production of certifiable systems once installed (for tool qualification refer to ED124/DO-297) [ED 07].

A1.2.2. *Means: an environment for modeling and simulating*

To carry out a development process according to the PMM approach, a project team should have a modeling and simulation environment.

This environment may be logically divided into two parts, the front-end and the back-end, even if it can be a physically integrated environment.

The first part, the front-end, allows the specifier and designer to develop specification and design descriptions models. It also allows system models to be configured from specifications and design descriptions previously modeled as building blocks.

It also allows the validator and the verifier to prepare validation processes (of specifications) and verification processes of different developmental steps (the design verification as first step), according to two axes (1) by designing a validation bench (respectively, a verification bench) and (2) by selecting the validation cases (respectively, verification cases) and the validation scenarios (respectively, verification cases).

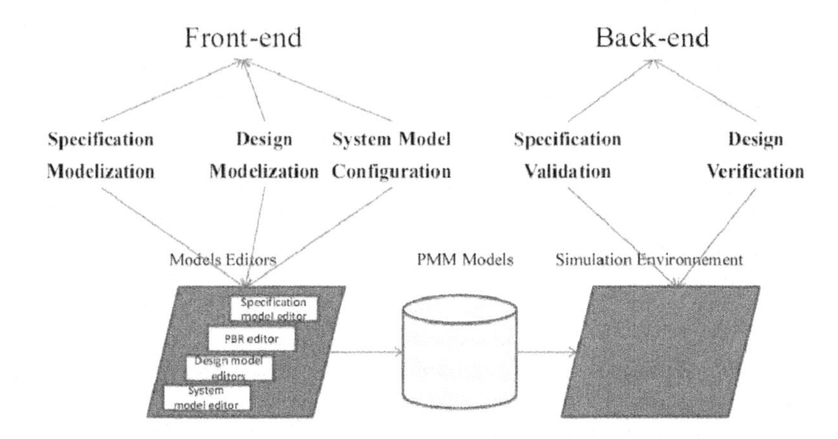

Figure A1.1. *PMM workbench conceptual design. For a color version of this figure, see www.iste.co.uk/micouin/MBSE.zip*

The second part of the development environment, back-end, allows the validator and verifier to validate specification model and verify design models.

Many simulation environments on the market, based on simulation languages such as VHDM-AMS or even Modelica, may implement the back-end of PMM methodology; however, we do not know of an environment on the market capable of implementing the front-end, given the novelty of concepts such as those of the specification model or PBRs.

It is, therefore, a characterization of this front-end, which we will focus on below.

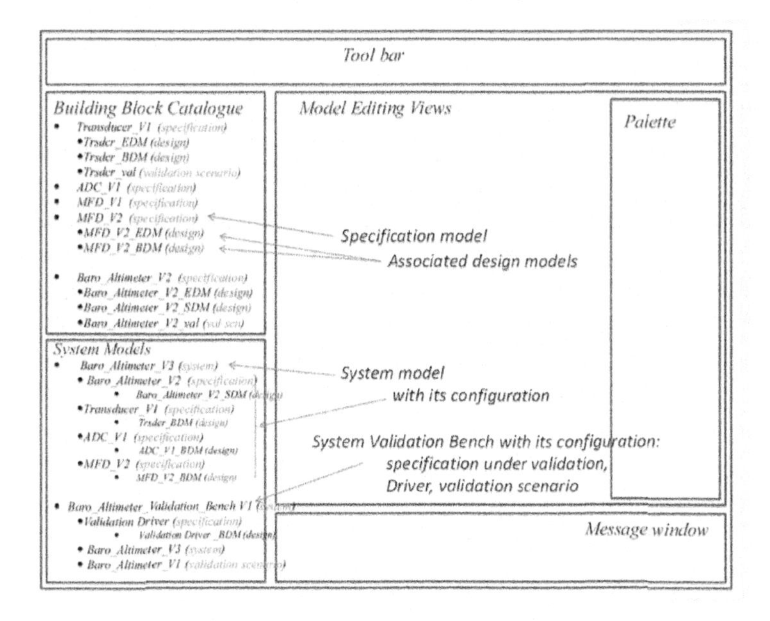

Figure A1.2. *Front-end main window and views on PMM models. For a color version of this figure, see www.iste.co.uk/micouin/MBSE.zip*

Figure A1.2 gives one possible view of the main editor, as well as a view on the database of PMM models (middle part of Figure A1.1) on the left part of the editor (Figure A1.2).

The left part of this main editor gives a view on the two categories of objects:

– on the top, a catalog of building blocks;

– on the bottom, system models and validation and verification bench models.

In the catalog of building blocks, the following are listed: (1) specification models and (2) design models associated with different types (EDM for equation models, SDM for structural models, BDM for behavioral models and RDM for reliability models). For example, we can see that there may be several versions of a specification model (for example,

MFD_V1 and MFD_V2). The design models are secondary entities relatively to a specification model, which they are attached. Design models are associated with this specification model, such as an equation design model (EDM) that validates and exactifies the specification model from which it originates. Also associated are a number of behavioral design models (BDMs) that describe different architectures and behavioral solutions and even a number of structural design models (SDMs) representing different architectures and structural solutions, to continue to decompose the model. We can then associate reliability models with these structural models to assess the conformity of design solutions to safety requirements. A specification model, associated with a set of design models, defines several variants of a building block.

In the bottom part, the following are listed: (1) SMs, with for each, the constituting building blocks, as well as (2) validation and verification bench models (which are only specific SMs with which data are associated: scenarios and validation or verification cases listed in files, execution traces of validation or verification scenarios).

A1.3. Producing a specification model (SSM)

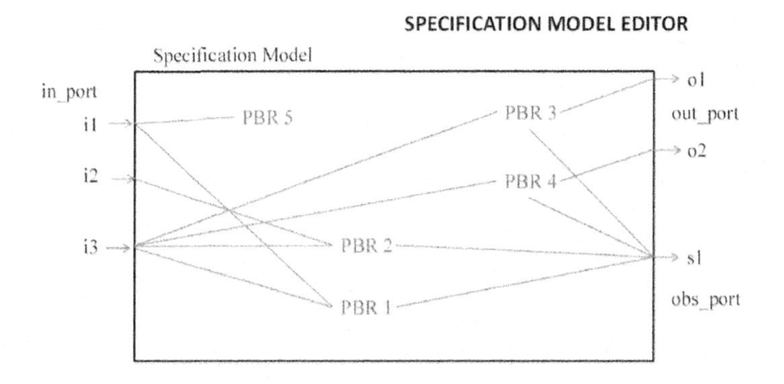

Figure A1.3. *PMM specification view*

Specifying the requirements and assumptions assigned to a type of systems by a specifier is carried out using a PMM editor, which has a type of systems associating a certain number of characteristics among which are found:

1) ports (i.e. interfaces between the specification model and its environment);

2) links between ports (i.e. PBRs or assumptions).

A1.3.1. *Ports*

The ports of a specification model are of three types: they may be input ports, output ports or even observable ports. In the logic of PMM (which is also that of block diagrams), an input represents a cause, whereas an output represents a visible effect of a type of modeled systems. Finally, an observable is a probe that the specifier uses to observe internal states of the SM.

Moreover, ports have a nature and a type. A port may be of a continuous nature or of a discrete nature. This allows us to characterize the quantities that flow through these ports. Thus, ports allowing quantities related to a fuel flow, crossing the borders of a fuel system, may be considered as continuous ports, while electronic signals will pass through discrete ports. In addition, ports shall be typed. Thus, continuous ports are either of the real type or a composite type where each component is real. Discrete ports have a more extensive typology. They may be scalar (enumerative, integer or real) or composite (homogenous: arrays, or heterogeneous: records).

If, for example, we consider a fuel system, then the fuel flow rate, required by the engine(s) at a given moment, may form a continuous input (i.e. a cause) of a fuel system, whereas the pressure at which the fuel is supplied to the engine(s) may be a continuous output (i.e. an effect) of the fuel system to respond to the engine demand. The specifier may also introduce an observable that helps us know the actual fuel quantity contained in the fuel system. However, this observable will not be confused with a fuel level indication (required by the regulation) that should be processed as an output, since it corresponds to an effective output, not only of the model, but also of the physical corresponding systems (whereas an observable does not correspond necessarly to the output of a system).

A1.3.2. *Property-based requirements and assumptions*

The PBRs of a specification model are declarations of links between the ports of a specification model. A PBR can link one or many ports together. We can subdivide these PBRs into two classes, assumptions and system requirements:

– *Assumptions*: PBRs that link inputs together are assumptions. As examples, we can assume that the quantities supplied by the environment to a system are always situated between a minimum m and a maximum M value. We can also assume that two consecutive events at an input port are separated by at least a time interval Δt, or even that a particular activation order of n input ports should be respected by the system environment. These are assumptions, since respecting the input constraints of a system is not the responsibility of the system itself. Nevertheless, these assumptions shall be validated.

– *System requirements:* the validation of these assumptions being successful, the system will function correctly and completely in the whole domain delimited by these assumptions. This operating domain of the system is determined by the PBRs other than the assumptions, i.e.

system requirements. A system requirement is a PBR that links one or many ports, of which at least one is an output port or an observable.

Each PBR will not only be declared, but also defined, that is to say it will be expressed in a non-ambiguous language, which can be compiled (after an eventual transformation) and whose compilation result can be executed in a simulation environment. For this, there are three possibilities: (1) to define a specific PBR modeling language with the required properties, (2) to directly use the simulation language to express the PBRs and (3) to provide a (graphical) support allowing the PBRs to be defined.

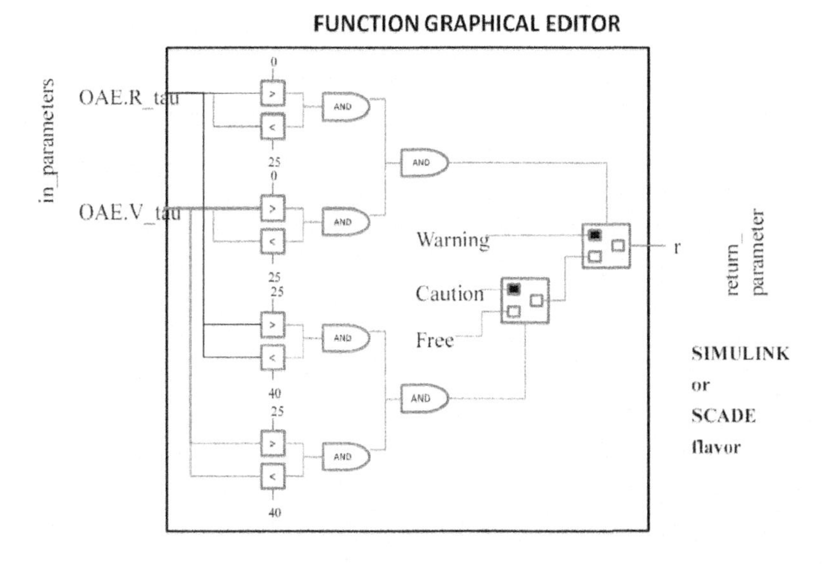

Figure A1.4. *Function and PBR graphical editor*

Figure A1.4 represents a means of defining a PBR (and more generally, a function) using a graphical editor inspired by graphical editors such as those provided by SIMULINK or even by SCADE. However, specifiers not put off by a textual formalism should be able to directly access a PBR model, either to create it or to modify it (particularly during the PBR validation phases). Figure A1.4 is a graphical representation of the definition of a function. For the definition of a PBR (that is only ever a Boolean function), the function inputs correspond to the ports that are linked to each other by PBR in the declaration at the specification model level. The connection between inputs and outputs is established by using different types of connectors (logical, algebraic, comparative, operational conditional and iterative, functional, temporal, etc.) that link them to varying or constant quantities. These graphical connectors are available and accessible via palettes.

A1.4. Producing design models

Producing design solutions associated with a specification model may be carried out in four different ways. The first involves producing an EDM from the specification model. The second involves developing a BDM that implements a behavior model for the type of systems. The third involves the decomposition of a system into an SDM composed of a certain number of components linked together and to ports of the SM by links. Finally to carry out reliability studies, the fourth involves designing RDMs. As there is no formal difference between an SDM and an RDM, the latter will not be discussed below.

A1.4.1. *Equation design models*

EDMs were introduced in section 7.4.2. An EDM is a design model whose purpose is to allow the validation of a specification model. It is, therefore, based on a transformation that moves from a PBR to its dual: a design equation.

The schema for generating design equations is based on the fact that the most general form of a PBR is an expression.

Req: when C => O.P ∈ D.

An equation derived from this PBR will, therefore, be in the following form:

Eq: O.P:= x ∈ D when C.

In other words, assign to the property O.P any value *x* (that can be chosen randomly if needed) belonging to domain D when condition C is verified.

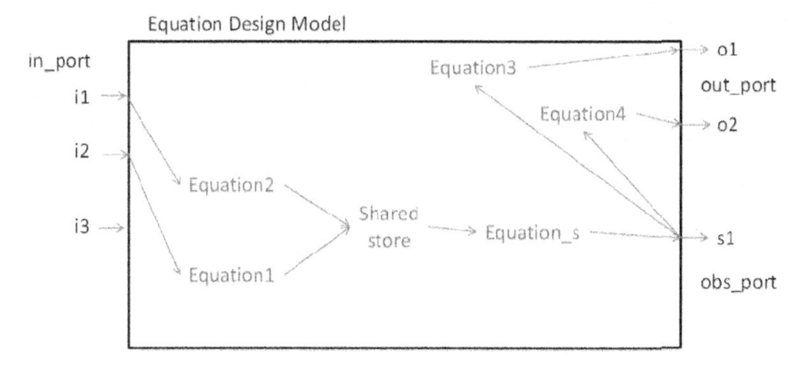

Figure A1.5. *Equation design model view*

However, some PBRs may not give rise to any equation in the EDM model; this is the case for assumptions, as they are not produced by the system but by the environment. This is also the case for all coherence controls that may be defined by outputs or observables (for example, the fuel pressure at the output of a fuel system will not exceed 3.0 bars) and that are nicely expressed in the form of a PBR.

Other PBRs may require a slightly more complicated generation schema than the base schema shown above. In particular, this is the case when two or more inputs can independently affect an output or an observable. In this situation, several input flows (of information, matter or energy) converge toward the same system output. In this case, it may be necessary to control the access conditions of flows entering a shared resource, which itself feeds the output. This is what is illustrated by the EDM in Figure A1.5 on which two equations (equations 1 and 2) converge both on the same shared resource.

A1.4.2. *Behavioral design model*

As mentioned in section 7.4.1, the second type of behavioral model (alongside equation models) is provided by the BDM.

A BDM is developed, by a designer, based on a validated specification model.

Figure A1.6 illustrates how this can be a graphical representation of a BDM.

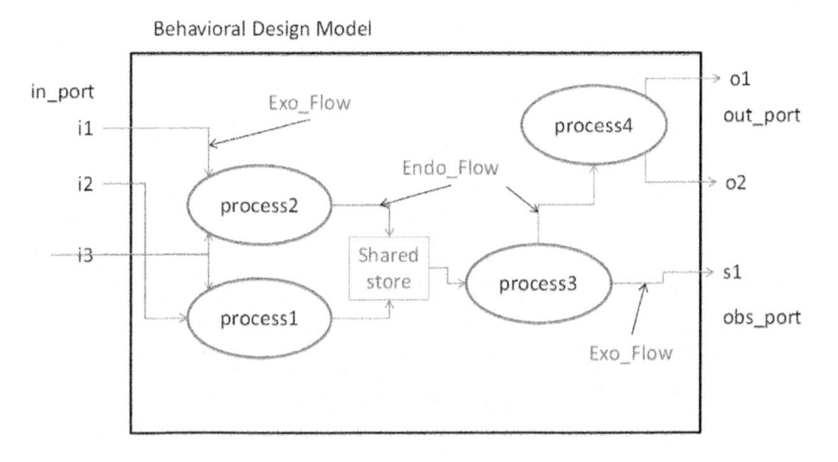

Figure A1.6. *Behavioral design model view*

This figure shows the following features:

1) one or more processes;

2) from zero to many shared resources;

3) a network of flows.

These features of a BDM can be declared and interconnected using graphical patterns selected on a palette.

The processes of a BDM have ports defined in the same way as ports of a specification (nature, type and direction). These ports link processes to external ports, to other processes or even to shared resources by flows. These links formed by flows will be consistent in nature, type and direction for ports and in nature and type for flows.

The processes may be discrete, continuous or hybrid:

1) In the first case (discrete processes), they do not have ports of a continuous nature.

2) In the second case (continuous processes), they do not have ports of a discrete nature.

3) In the third case (hybrid processes), they mix discrete and continuous ports.

Shared resources allow discrete or hybrid processes to consistently synchronize and exchange quantities, as required, by operators accessing resources.

Flows connect either ports of the specification model to internal processes (external flows), or internal processes between them (internal flows), or internal processes to shared resources (and *vice versa*). Note that a flow can have several destinations but one single source.

Once declared (using a BDM editor), the processes composing such a model will be defined.

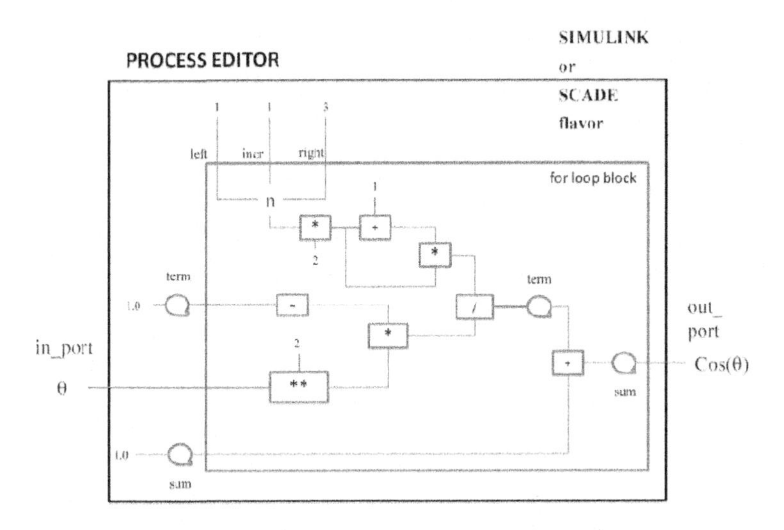

Figure A1.7. *Process graphical editor*

As with PBR, each process will not only be declared (using the BDM editor) but also be defined, that is to say, it will be expressed in a non-ambiguous language, which can be compiled (after an eventual transformation) and whose compilation result can be executed in a simulation environment. For this, there are three options: (1) to define a specific modeling language for the process description, which has the required properties, (2) to directly use the simulation language features to define these processes and (3) to provide a (graphical) support allowing the processes to be defined.

Figure A1.7 shows how to define a process using a graphical editor inspired by graphical editors such as those provided by SIMULINK or even by SCADE. However, designers not put off by a textual formalism should be able to directly access a process model, either to create it or to modify it (particularly in the verification phases of the design model).

This figure is a graphical representation of the definition of a process computing cosine of an angle θ using a MacLaurin series up to order 6. To define a process, the process input ports are linked to obsrevable or output ports using different types of connectors (logical, algebraic, comparative, operational conditional and iterative, functional, temporal, etc.) which also connect them to (1) constant or varying quantities, local to the process, or eventually to (2) shared resources. These connectors are available and accessible via palettes.

A1.4.3. *Structural design model*

As mentioned in section 7.6, the PMM methodology defines two kinds of static model. The first kind includes SDMs, while the second kind is described in the next section.

An SDM is developed, by a designer, based on a validated specification model. These design models with their characteristics were discussed in section 7.6.2.

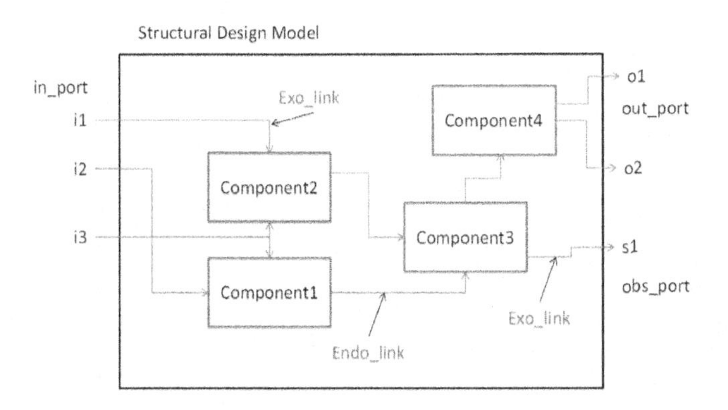

Figure A1.8. *Structural design model view*

Figure A1.8 provides an illustration of a graphical representation of an SDM.

This figure shows:

1) one to several components;

2) a network of links.

The characteristic features of an SDM can be declared and interconnected using models selected on palettes.

The components of an SDM have ports defined as with specification ports (nature, type and direction). These ports allow us to connect the component to external ports, to other components by means of links. These links will be consistent in nature, type and direction for ports and in nature and type for links.

The links connect either ports of the specification model to component ports – these are external links –, or component ports together – these are internal links. Note that a link may have several destinations but one single source.

Components of an SDM are receptors designed to receive building blocks; therefore, the specification model has exactly the same ports as those of the receptor component. A component is, therefore, a kind of placeholder designed to be linked to a building block during the configuration of an system model (SM) (see section A1.5).

A1.4.4. *Reliability design model*

In PMM, the second type of structural model is the reliability design model (RDM).

An RDM is developed, by a designer, in reference to an SDM. These reliability models and their characteristics were presented in section 10.3.6.

These models do not introduce additional syntactical elements with regard to an SDM editor.

A1.5. *Producing a system model (SM)*

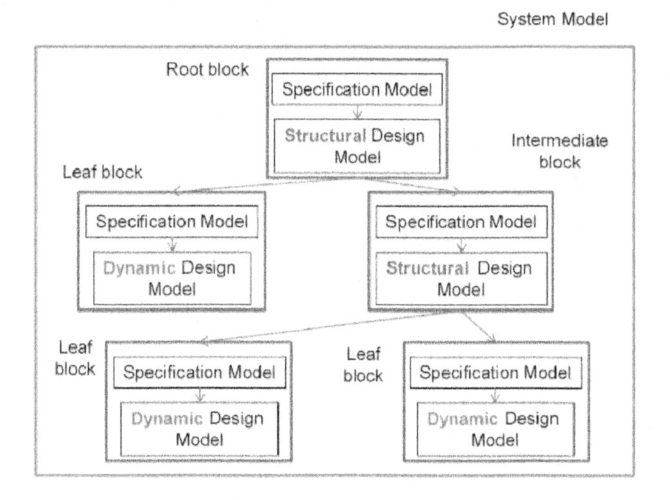

Figure A1.9. *System model and specification and design tree. For a color version of this figure, see www.iste.co.uk/micouin/MBSE.zip*

In PMM, an SM is a particular assembly (a configuration) of a set of building blocks, structured like a tree.

An SM is, therefore, composed of a root block linked to zero or at least two successor blocks.

If the root block is without a successor, then it is ultimately composed of a specification model and a behavioral model (EDM or BDM).

Whatever is true for the root block (without successors), is also true for any leaf block (i.e. without successors). A leaf block is composed of a specification model and a behavioral model (EDM or BDM). The behavioral model associated with the specification model of a leaf block is one of the design models associated with this specification and is available in the block catalog. Therefore, all that is required is to select it using a “drag & drop” operation in the SM editor.

With regard to an intermediate block (it has successors) or the root block (if it has successors), such blocks are inevitably composed of a specification model and a structural model (SDM or RDM).

As these structural models contain at least two components[1], it is therefore necessary to link these components with building blocks residing in the building block catalog, as shown in Figure A1.10.

In this example, the SM Robust_Pressure_Altimeter consists of a specification model (Robust_Pressure_ Altimeter) and an SDM (Robust_Pressure_Altimeter_SDM). This design model includes. The_Back_up_ADC (the other components are not mentioned). The latter is linked (binding) to the building block formed by the specification model Air_Data_Computer and an associated equation design model (EDM), i.e. Air_Data_Computer_EDM.

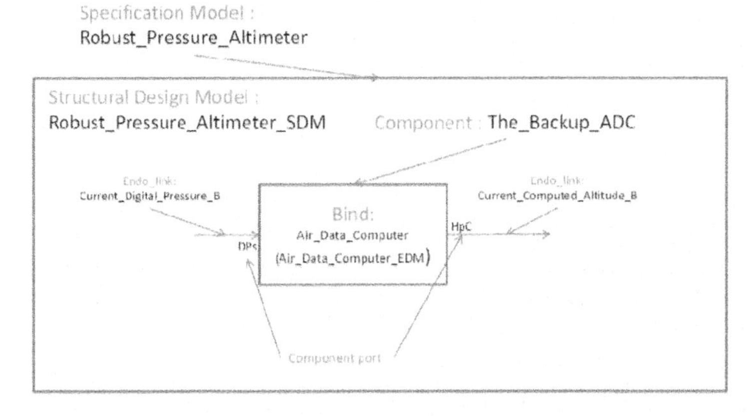

Figure A1.10. *Component-block binding in system model editor. For a color version of this figure, see www.iste.co.uk/micouin/MBSE.zip*

1 An SDM consisting of a single component is a meaningless construct.

Bibliography

[ABR 05] ABRIAL J.R., *The B-Book: Assigning Programs to Meanings*, Cambridge University Press, November 3, 2005.

[ALA 60] ALAIN E.C., *Humanités 1925, Les idées et les âges, Les Passions et la Sagesse*, Gallimard, La Pléiade, 1960.

[ANS 03] ANSI/EIA632, Processes for engineering a system, GEIA, Arlington, VA, 2003.

[ARE 61] ARENDT H., "It is as though I had the right to call the heel of my shoe a hammer because I, like most women, use it to drive nails into the wall", *What is Authority?, in between Past and Future: Six Exercises in Political Thought*, The Viking Press, New York, 1961.

[ARI 24]ARISTOTLE, *Metaphysics Book Z*, Chapter III, translated by W.D. Ross, Oxford University Press, December 31, 1924.

[ARI 99] ARISTOTLE, *Nichomachean Ethics*, translated by W.D. Ross, Batoche Books, 1999.

[ASH 03] ASHENDEN P.J., PETERSON G.D., TEEGARDEN D.A., *The System Designer's Guide to VHDL-AMS*, MK Publishers, 2003.

[AUV 09] AUVINEN P., Achievement of intersubjectivity in airline cockpit interaction, Academic dissertation, University of Tampere, 2009.

[BER 69] VON BERTALANFFY L., *General System Theory, Bertalanffy*, George Braziller Inc., New York, 1969.

[BIN 76] BINDRA D.A., *Theory of Intelligent Behavior*, Wiley and Sons, 1976.

[BKC 13] BKCASE Editorial Board, SEBOK: guide to the systems engineering body of knowledge, 14 November 2013. http://www.sebokwiki.org/wiki/Guide_to_the_Systems_Engineering_Body_of_Knowledge_%28SEBoK%29.

[BOU 14] BOULANGER J.L., FORNARI F.X., DION B., SCADE: Language and Application, ISTE, forthcoming.

[BOU 56] BOURBAKI N., *Eléments de mathématiques*, part I, Livre I, Chapter III, Pre-ordred sets, Hermann & Cie, 1956.

[BOU 14] BOURQUE P., FAIRLEY R.E. (eds.), *Guide to the Software Engineering Body of Knowledge, Version 3.0*, IEEE Computer Society, 2014. Available at http://www.computer.org/portal/web/swebok/swebokv3pdf.

[BRO 84] BROWN J.S., KLEER DE J., "A FrameWork for qualitative Physics, Xerox", *Proceedings of the Cognitive Science Society,* sixth annual, 1984.

[BSE 14] BS-EN 1325, Value Management, Value Analysis, Functional Analysis Vocabulary, Value Analysis and Functional Analysis, 2014.

[BUN 73] BUNGE M.A., *Method, Model and Matter*, Reidel, New York, 1973.

[BUN 77] BUNGE M.A., *Ontology I: The Furniture of the World,* Treatise of Basic Philosophy, Reidel, New York, vol. 3, 1977.

[BUN 77] BUNGE M.A., SANGALLI A., "A theory of properties and kinds", *International Journal of General Systems*, vol. 3, pp. 183–190, 1977.

[BUN 79] BUNGE M.A., *Ontology II: A World of Systems,* Treatise on Basic Philosophy, Reidel, New York, vol. 4, 1979.

[BUN 83a] BUNGE M.A., *Epsitemology & Methodology I: Exploring the World,* Treatise of Basic Philosophy, Reidel, New York, vol. 5, 1983.

[BUN 83b] BUNGE M.A., *Epistemology and Methodology II: Understanding the World,* Treatise of Basic Philosophy, Reidel, New York, vol. 6, 1983.

[BUN 03] BUNGE M.A., *Emergence and Convergence: Qualitative Novelty and the Unity of Knowledge* (Toronto Studies in Philosophy), University of Toronto Press, 2003.

[BUN 10] BUNGE M.A., *Matter and Mind,* Boston Studies in Philosophy of Sciences, Springer, vol. 287, 2010.

[BUN 12] BUNGE M.A., *Evaluating Philosophies,* Boston Studies in Philosophy of Sciences, Springer, vol. 295, 2012.

[CHA 97] CHANGEUX J.P., *Neuronal Man*, Princeton University Press, 1997.

[CHA 00] CHANDRASEKARAN B., JOSEPHSON J.R., "Function in Device Representation, Engineering with Computers", *Special Issue on Computer Aided Engineering,* 2000.

[CHU 61] CHURCHMAN C.W., ACKOFF R.L., ARNOFF E.L., *Introduction to Operations Research*, John Wiley & Sons, 1961.

[DES 07] DESCARTES René, *Discourse on Method*, Part 2, p. 9, translated by Jonathan Bennett, 2007, http://www.earlymodern texts.com/.

[DIL 90] DILLINGER M., "On the concept of a language", in WEINGARTNER P., DORN G.J.W. (eds.), *Studies on Mario Bunge Treatise*, Rodopi, pp. 5–26, 1990.

[DOR 02] DORI D., *Object Process Methodology, A Holistic System Paradigm*, Springer, pp. 251–252, 2002.

[DUR 02] DURAND D., *La systèmique*, PUF, 2002.

[EAS 12a] EASA CS25, Certification specifications and acceptable means of compliance for large aeroplanes, Amendment 14, 12 July 2012.

[EAS 12b] EASA CS29, Certification specifications for large rotorcraft, Amendment 3, 11 December 2012.

[ED 00] ED-80/DO-254 Design assurance guidance for airborne electronic hardware, EUROCAE, RTCA, 19 April 2000.

[ED 07] ED124/DO-297, Integrated modular avionics (IMA) development guidance and certification considerations, EUROCAE, June 2007.

[ED 12] ED-12C/DO-178C, Software considerations in airborne systems and equipment certification, EUROCAE, RTCA, 1 January 2012.

[ENG 98] ENGESTROM Y., MIDDLETON D. (eds.), *Cognition and Communication at Work*, Cambridge University Press, 1998.

[FAR 10] FAR Part 29 Airworthiness Standards, Transport category Rotorcraft, FAA latest version, 1 January 2010.

[GAZ 09] GAZZANIGA M.S., *The Cognitive Neurosciences*, 4th ed., MIT, 2009.

[GER 90] GERO J.S., "Design prototypes: a knowledge representation schema for design", *AI Magazine*, vol. 11, no. 4, pp. 26–36, 1990.

[GRA 98] GRADY J.R., *System Requirements Analysis*, McGraw Hill, 1998.

[HAL 62] HALL A.D., *A Methodology for Systems Engineering*, Van Nostrand, Princeton, New Jersey, 1962.

[HAR 98] HAREL D., POLITI M., *Modeling Reactive Systems with Statecharts: The STATEMATE Approach*, McGraw-Hill, 1998.

[HAT 03] HATCHUEL A., WEIL B., "A new approach of innovative design: an introduction to C-K theory", *International Conference on Engineering Design*, 19–21 August, 2003.

[HEB 49] HEBB D.O., *The Organization of Behavior: A Neuropsychological Theory*, Wiley and Sons, New York, 1949.

[HUB 84] HUBKA W., EDER W.E., *Theory of Technical Systems: A Total Concept Theory for Engineering Design*, Springer-Verlag, 1984.

[HUT 96] HUTCHINS E., *Cognition in the Wild*, Bradford Books, 1996.

[ICA 64] ICAO, *Manual of the ICAO Standard Atmosphere Extended to 32 Kilometers*, 2nd ed., International Civil Aviation Organization, Montreal, 1964.

[IEE 05] IEEE Standard 1220-2005, *IEEE Standard for Application and Management of the Systems Engineering Process*, IEEE Computer Society, 9 September 2005.

[IEE 07] *IEEE Standard VHDL Analog and Mixed-Signal Extensions*, IEEE 1076-1, IEEE Computer Society, 2007.

[IEE 08] *IEEE Standard VHDL Language Reference Manual*, IEEE 1076, IEEE Computer Society, 2008.

[INC 07] "INCOSE, Systems Engineering Vision 2020", Document INCOSE-TP-2004-004-02, Version 2.03, September 2007. Available at https://www.incose.org/ProductsPubs/pdf/SEVision 2020_20071003_v2_03.pdf.

[ISO 05] ISO/IEC 15288, Systems Engineering – System Life Cycle, Geneva, Switzerland, 2005.

[ISO 11] ISO/IEC/IEEE 29148: 2011 Systems and Software Engineering – Life Cycle Processes – Requirements Engineering, ISO Geneva, Switzerland, 1 December 2011.

[KAR 06] KARNOPP D.C., MARGOLIS D.L., ROSENBERG R.C., *Systems Dynamics, Modeling and Simulation of Mechatronic Systems*, John Wiley & Sons, 2006.

[KRE 00] KREIMAN G., KOCH C., FRIED I., "Imagery neurons in the human brain", *Nature*, vol. 408, pp. 357–361, 16 November 2000.

[KUH 70] KUHN T., *The Structure of Scientific Revolutions*, (2nd revised ed.), University of Chicago Press, 1 April 1970.

[LAS 50] LASLEY K., "In search of the engram", *Society of Experimental Biology Symposium*, vol. 4, pp. 454–482, 1950.

[LE 84] LE MOIGNE J.-L., *La théorie du système générale*, PUF, 1984.

[LET 02] LETIER E., VAN LAMSWERDE A., "Deriving operational software specification from system goals", *Proceedings of the 10th ACM SIGSOFT Symposium on Foundations of Software Engineering 2002*, Charleston, South Carolina, SC, 18–22 November, 2002.

[LUR 76] LURIA A., *The Working Brain*, Basic Books, 1976.

[LUZ 10] LUZEAUX D., RUAULT J.-R. (eds.), *Systems of Systems*, ISTE Ltd and John Wiley & Sons, 2010.

[MAH 97] MAHNER M., BUNGE M.A., *Foundations of Biophilosophy*, Springer, pp. 367–376, 1997.

[MCD 03] MCDERMID J., NICHOLSON M., *Extending PSSA for Complex Systems*, ISSC, Ottawa, August 2003.

[MER 42] MERTON R.K., *The Sociology of Science*, University of Chicago Press, 1942.

[MIC 29] MICHAUX H., "I built myself on a missing column (*Je me suis bâti sur une colonne absente)*", *je suis né troué*, Ecuador, 1929.

[MIC 01] MICHAUX H., *Ecuador: A Travel Journal*, Northwestern University Press, Reprint, 2001.

[MIC 06] MICOUIN P., KIEFFER J.P., *Place and forms of decision-making in systems engineering (in French)*, RNTI-E-8, Cepadues Editions, pp. 37–65, 2006.

[MIC 08] MICOUIN P., "Toward a property based requirements theory: system requirements structured as a semilattice", *Systems Engineering*, vol. 11, no. 3, pp. 235–245, 2008.

[MIC 13] MICOUIN P., "Model based systems engineering using VHDL-AMS", *Conference on Systems Engineering Research*, Procedia Computer Science, Elsevier, vol. 16, pp. 128–137, 2013.

[MIC 14] MICOUIN P., "Property-model methodology: a model-based systems engineering approach using VHDL-AMS", *Systems Engineering*, vol. 17, no. 3, pp. 249–263, 2014.

[MOD 10] MODARES M., KAMINSKIY M., KRISTOV V., *Reliability Engineering and Risk Analysis, A Practical Guide*, CRC Press, Boca Raton, FL, 2010.

[MOD 12] MODELICA Association, "Modelica – a unified object-oriented language for systems modeling, language specification", Version 3.3, 9 May 2012. Available at https://modelica.org/documents/ModelicaSpec33.pdf.

[MOR 11] MORIARTY P., "Science as a public good", 2011. http://www.bloomsburyacademic.com/view/A-Manifesto-for-the-Public-University/chapter-ba-9781849666459-chapter-005.xml.

[OMG 12] OMG Systems Modeling Language (OMG SysML™), Version 1.3, June 2012.

[PAL 14] PALM W. III, *System Dynamics*, 3rd ed., McGraw-Hill International Edition, 2014.

[POL 45] POLYA G., *How to Solve it?*, Princeton University Press, 1945.

[POP 72] POPPER K.R., *Objective Knowledge: An Evolutionary Approach*, Oxford University Press, New York, 1972.

[POP 02] POPPER K.R, *Conjectures and Refutations. The Growth of Scientific Knowledge*, Routledge Classics, 2002.

[RAY 03] RAYNAUD D., Sociologie des controverses scientifiques [Sociology of scientific controversies], PUF, 2003.

[RIZ 05] RIZZOLATI G., "The mirror neuron system and its function in humans", *Anatomy and Embryology*, vol. 210, nos. 5–6, pp. 419–421, December 2005.

[ROO 91] ROOZENBURG N.J.M., EEKELS J., *Product Design: Fundamentals & Methods*, John Wiley & Sons, 1991.

[RUS 13] RUSSEL S.J., NORVIG P., *Artificial Intelligence: A Modern Approach*, Pearson, 2013.

[SAE 96] SAE-AS8002A, Air Data Computer: Minimum Performance *Standard*, Aerospace Standard SAE International, 1996–09.

[SAE 09] SAE-AS5506A, Architecture Analysis & Design Language (AADL), January 2009.

[SAE 10] SAE-ARP4754A, Guidelines for Development of Civil Aircraft and Systems, SAE, 2010.

[SAU 00] SAUSSURE DE F., *Cours de Linguistique Générale*, Payot, 1900.

[SCH 83] SCHON D.A., *The Reflective Practitioner, How Professionals Think in Action*, Basic Books, 1983.

[SCO 97] SCOTT J., *Systems Engineering for Commercial Aircraft*, Ashgate Publishing Company, 1997.

[SIM 91] SIMON H.A., "Simon, Bounded Rationality and Organizational Learning", *Organization Science*, vol. 2, no. 1, pp. 125–134, May 1991.

[SPE 07] SPELKE E.S., KINZLER K.D., "Core knowledge", *Developmental Science*, vol. 10, no. 1, pp. 89–96, 2007.

[SUH 05] SUH N.P., *Complexity, Theory and Application*, Oxford University Press, 2005.

[THO 75] THOMA J.U., *Introduction to Bond Graphs and their Applications*, Pergamon Press, 1975.

[VIT 14] VITRUVIUS, *De architectura, The Ten Books on Architecture*, translated by Morris Hicky Morgan, Harvard University Press, Cambridge, 1914.

[WAR 86] WARD P.T., MELLOR S.J., *Structured Development for Real-Time Systems*, Prentice Hall, 1986.

[WAR 96] WARRENDALE P.A., Guidelines and Methods for Conducting the Safety Assessment Process on Civil Airborne Systems and Equipment, SAE, SAE-ARP4761, 1996.

[WYM 93] WYMORE A.W., *Model-Based Systems Engineering*, CRC Press, 1993.

Index

A, B

ARP4754A, 46
assertion, 121, 124, 163
assumption, 11, 15, 42, 48, 52, 90, 104, 118, 124, 125, 151, 153, 154, 172, 196, 222, 228
  semantic, 65
behavior, 4, 20, 22, 23, 25, 28, 30–34, 36, 38, 40, 66, 68–70, 72, 75, 85, 109, 133, 136, 139, 159, 160, 173, 191, 192, 197, 218
  actual, 32
  expected, 38
  observable, 66
  really possible, 33
  the, 70
  theoritical model, 70
belief, 56, 57

C

concept, 4, 7, 12, 13, 19, 23, 30, 31, 34, 36, 45, 46, 59–62, 65, 80, 97, 102, 105, 110, 119, 128, 151, 153, 156, 169, 190, 197, 199, 214, 215, 219
  designation, 60
  meaning, 13, 61, 80
configuration management, 82, 92, 93, 213
contract theorem, 98, 181, 233

D

design, 16, 24, 28–30, 32, 33, 37, 56, 58, 69, 83–87, 89, 90, 92, 95, 96, 100, 102, 105, 107, 108, 125, 127–135, 139, 140, 142–149, 158–160, 162, 173–179, 182, 184, 187–190, 200–202, 204–208, 211, 212, 216, 218, 226–233
  choice, 56, 89, 129, 201
  error, 125, 174–176, 178, 179, 229
  model, 92, 102, 131–135, 139, 140, 142, 143, 145, 146, 159, 160, 162, 173–179, 182, 204–208, 212, 218, 227, 229, 230, 233

a goal, 30
of means, 30
process, 29, 83, 84, 87, 89, 92, 95, 127, 130, 148, 149, 190, 201, 211, 212, 226, 229
reasoning, 56, 58
reliability design model (RDM), 207
solution, 86, 87, 90, 92, 96, 128, 139, 189, 228, 229
verification, 173, 176, 179, 182, 211
development, 26, 45, 79–84, 88, 90–93, 96, 102, 105, 121, 129, 153, 154, 156, 159, 167, 176, 181, 183, 184, 187, 206, 211, 213, 215, 216, 218, 233
error, 187

E

EASA, 84, 115, 184
Certification Review Item, 115
ED12C/DO-178C, 121
ED79A/ARP4754A, 32, 79–81, 88, 91–93, 99, 100, 106, 170, 188
EIA, 79, 89, 96, 99, 100, 129, 148, 180, 194, 200, 211, 212
EIA632, 79, 80, 83, 85, 87, 148
engineering, 4, 23–28, 30, 33, 35, 40–42, 45, 48, 55, 75, 79–82, 84, 86, 91, 96, 97, 101, 102, 110, 129, 152, 156, 165, 169, 181, 183, 184
MBSE, 101
process, 23, 25, 79, 81, 82
systems, 80, 81
error, 25, 28, 37, 38, 110, 116, 125, 157, 164, 171, 174, 180, 191, 193, 220, 232, 233

F

FAA, 30, 83, 115, 184
issue paper, 115
fact, 4, 5, 9, 13, 18, 20, 22–24, 31, 34, 42, 44, 46, 49, 55, 56, 63, 66, 80, 101, 106, 112, 122, 149, 161, 162, 165, 209
statement, 23
failure, 25, 28, 37–39, 185, 106, 154, 170, 172, 173 186–199, 203–207, 209–211, 217, 219, 220
tree, 73, 207
fault, 25, 38, 73, 74, 90, 191, 192, 200, 207, 208
faulty state, 38
function, 25, 30–38, 101, 107, 122, 124, 125, 138, 159, 163–165, 177, 188, 191, 192, 194, 197, 198, 205, 206, 219, 220, 223
according to ARP4754A, 32
analysis, 34
Behavior-Structure, 32, 39, 191

G, I

genders, 67, 69
species and types, 13, 103, 105, 111, 112
general system theory (GST), 3
IEEE, 100, 101, 106, 107, 108, 215
IEEE1220, 101

implementation, 32, 56, 81, 82, 91–93, 129, 169–175, 179–182, 191
  verification, 170, 172
installation, 30, 139, 140, 173–176, 179, 182, 184, 187, 227, 232, 233
integration, 82, 88, 89, 92, 148, 158, 171, 173, 174, 176, 178, 179, 181, 182, 187, 216, 231–233
ISO, 24, 25, 79, 106–108
ISO15288, 24, 25, 27, 79, 80, 100

K, L

knowledge, 4, 12, 23–25, 41–45, 47, 48, 50, 51, 56, 57, 59, 80, 99, 114, 118, 135, 153, 165, 227, 229, 230
  applied science, 51
  body of, 46, 50
  conceptual, 44, 45
  falsifiable, 47
  intersubjective, 44, 48, 114
  objective, 42, 47, 48, 50, 114
  procedural, 45
  scientific, 41
  subjective, 42, 43
  system, 6, 23, 42, 56, 59, 80
  technological, 50, 56, 80, 99
  truth, 47, 99
  unverifiable, 48
language, 10, 16, 19, 52, 64, 70, 73, 74, 97, 101, 162, 165
  expressiveness, 71, 73
  SA-RT, 74
  SYSML, 74
  VHDL, 74
law, 15, 17, 21, 22, 28, 29, 35, 48, 51–53, 55, 56, 59, 65, 66, 69, 70, 138, 184, 204
  as property, 15
  statement, 18, 21, 22, 28, 52, 53, 59, 65, 66, 69, 70

M, N, O

mode, 11, 12, 25, 30, 36, 37, 103, 119, 120, 136, 138
  commutation, 36
model, 11, 42, 53–55, 59, 64, 66–73, 75, 81, 82, 96, 97, 102, 122, 131–136, 138–140, 142–145, 152, 155–157, 159–161, 167, 170, 171, 173, 176–178, 182, 184, 191, 198, 206–208, 212, 213, 217, 218, 219, 221, 225, 228–232
  object, 66, 67, 69, 71, 72
  representativeness, 71, 72, 73
  symbolic, 64
  theoretical, 66, 68, 69, 70, 72
no-function-in-structure, 33
object, 3–7, 9, 11, 12, 14–23, 32, 33, 44, 45, 49, 56, 59, 60, 64, 66–69, 70–75, 99, 103, 104, 110–112, 120, 161
  abstract, 11
  denotation, 60
ontology, 42

P

PBR, 97, 102–104, 106, 109–114, 117, 124, 130, 156, 163, 164, 196, 198, 199, 201, 209–211

assumption, 104
comparaison, 111
conjonction, 110, 112, 118, 123
objectivization, 102
obligation or prohibition, 104
type, 108
predicate, 46, 111, 125, 164
process, 4, 20, 22, 23, 29, 41, 44, 47, 51, 56, 58, 79, 80–83, 86–93, 95–97, 99, 100, 106, 107, 114, 115, 121, 125, 127, 129, 137–139, 141, 144, 146–148, 152–161, 163, 164, 167, 170–173, 178, 179, 184, 185, 187, 188, 191, 194, 195, 200, 201, 204, 211, 213, 215, 216, 218, 228, 231–233
production, 7, 74, 108, 141, 173, 174, 179, 182, 187, 216
property, 4, 11, 13–22, 25, 31, 56, 61, 63, 80, 96, 97, 102, 103, 109, 110, 119, 128, 131, 141, 152, 156, 167, 170, 173, 181, 184, 191, 193, 206, 209, 213, 221, 224
accidental, 13, 16
actualisation, 17
class, 13
concomitance, 14
conjonction, 14
disposition, 42
emergent, 4, 18
essential, 13, 20, 51
formal, 12, 13
generic, 20
intrinsic, 17
material, 11, 20
precedence, 14
quantitative, 20
specific, 20
structural, 4, 17
proposition, 12, 13, 22, 45–49, 51–53, 56, 57, 59, 63, 65, 66
meaning, 46
meaningful, 46
meaningless, 46
nomological, 22, 28, 51, 52, 65, 66
scientific, 56
truth, 13, 49
comparaison, 97
requirement, 46, 75, 84, 86, 89–91, 96–98, 100–108, 110, 111, 114–120, 124, 129, 130, 133, 136, 140, 141, 144, 147, 151, 153, 154, 156, 158, 162–166, 171, 172, 181, 184, 188, 190, 193–199, 201, 209–211, 228
derivation and assignation, 91, 97, 128
derivation, 86, 97, 130, 147
determination, 96, 188
source, 106, 108
system technical, 90
validation, 91, 92, 154, 158, 165, 161, 211

S

safety, 73, 79, 82, 84, 88–90, 96, 98, 108, 113, 115, 139, 154, 164, 167, 172, 183–191, 194, 196, 200–204, 206, 208, 209, 211, 214, 218
analysis, 188
ASA, 188, 190
CCA, 188, 191

CMA, 188
DAL, 203
fail safe concept, 190
FDAL, 188
FHA, 188
PASA, 188, 189, 200
PRA, 188
PSSA, 188, 200
quantitative requirement, 190
requirement, 84, 89, 90, 96, 184, 189, 190, 194, 196, 200, 201, 202, 203, 204, 206, 209, 211, 218
SSA, 188, 190
theorem, 208, 209
ZSA, 188
scenario and case, 96
semantical assumption, 52, 66, 69, 74
signs, 4, 10, 13, 16, 23, 24, 27, 42, 52, 59–62, 64–66, 68, 69
system, 4, 23, 24
simulation, 60, 66, 68, 70, 91, 92, 96, 97, 124, 125, 132, 134, 140, 143, 144, 158–167, 176, 177, 180, 182, 204, 206, 208, 214, 218, 219
bench, 96
space, 11
specification, 46, 58, 87, 95–102, 105, 113, 114, 121–125, 131, 132, 134–137, 139–145, 148, 149, 153, 157–163, 167, 169, 170, 173, 175, 177–179, 181, 182, 187, 197, 199, 200, 206, 208–211, 216–220, 223, 224, 226–233
choice, 56
error, 157
exacte, 98
impossible, 99
model, 96–98, 102, 121–125, 131, 132, 134–136, 139–145, 148, 157–163, 167, 173, 175, 177–179, 181, 182, 197, 199, 200, 206, 208, 217– 220, 224, 226, 227–233
over specification, 97, 99
specification model, 95, 96, 97, 98, 102, 121, 122
under specification, 97, 99
state, 4, 5, 8, 20–23, 25, 28, 29, 36, 37, 62, 68, 79, 80, 99, 103, 109, 122, 123, 135–137, 163, 173, 183, 192, 198, 209, 211, 220, 221, 223, 224, 227, 230
space, 15, 211
system, 3–10, 12, 14, 16, 18–21, 26–39, 43, 45, 47, 49, 52, 54, 59–62, 64, 65, 67, 68, 75, 79, 80, 82, 83, 84, 86–93, 95–98, 102–109, 119–124, 127–134, 136–148, 152, 154, 156–163, 166, 169–184, 188–190, 192–211, 213–216, 218, 219–223, 225, 226, 229, 231, 233
abstract, 3, 24, 45, 52, 61, 65
autotelic, 27, 51, 53
behavior, 4, 31, 75, 134
closed, 5
composition, 3, 4, 5, 39, 61, 68, 71, 95, 103, 104, 109, 123
concomitant effect, 25
concrete, 3, 7, 25

endo-structure, 4, 5, 8, 46, 61, 62, 63, 95, 123
environment state, 28
environment, 3, 4, 71
event, 4, 20, 103, 124
exo-structure, 4, 9, 46, 61, 63
failure, 25, 37, 106
fault, 25, 37
function, goal or teleology, 25, 27, 30, 32–34, 36, 41
heterotelic, 27, 31, 51
link, 8
mode, 36, 103, 131–133, 136, 140, 142–144, 159, 163, 175, 177, 178, 218, 231
of systems, 7
open, 5
state, 4, 20, 25, 36, 37, 103
structure, 3, 10, 19, 30–33, 39, 67, 71, 103, 104, 109, 190, 200, 208
subsystem level, 3, 9
subsystem, 9
technico-empirical, 27
technological, 4, 24, 28

T, V

TBR, 97, 101, 114, 115, 120
and MBSE, 101
implicit, 120
interpretation, 114
subjectivity, 100
truth, 13, 18, 22, 48, 49, 52, 53, 56, 57, 59, 63, 66, 163
approximate, 49, 52
formal, 49
partial, 49
validation, 44, 58, 80, 82, 90, 91, 93, 99, 139, 141, 151–172, 175, 179, 204, 211, 217, 218, 228
by analysis, 91, 92, 155, 172
by engineering judgment, 156
by modeling, 155, 172
by test, 156, 159
early, 91
effort and rigor, 91
means, 91
verification, 47, 48, 51, 54, 56, 58, 82, 91–93, 135, 139, 145, 148, 151, 169–182, 188, 189, 191, 206, 229, 231–233
by simulation, 177
device, 47, 48
goals, 170, 174
means, 92, 170
process, 47, 51, 58, 82, 92, 135, 145, 148, 151, 169, 170–174, 180, 188, 191, 206

Other titles from

in

Control, Systems and Industrial Engineering

2014

DAVIM Paulo J.
Machinability of Advanced Materials

FAVRE Bernard
Introduction to Sustainable Transports

MILLOT Patrick
Designing Human–Machine Cooperation Systems

OUSTALOUP Alain
Diversity and Non-integer Differentiation for System Dynamics

REZG Nidhal, DELLAGI Sofien, KHATAD Abdelhakim
Joint Optimization of Maintenance and Production Policies

2013

ALAZARD Daniel
Reverse Engineering in Control Design

ARIOUI Hichem, NEHAOUA Lamri
Driving Simulation

CHADLI Mohammed, COPPIER Hervé
Command-control for Real-time Systems

DAAFOUZ Jamal, TARBOURIECH Sophie, SIGALOTTI Mario
Hybrid Systems with Constraints

FEYEL Philippe
Loop-shaping Robust Control

FLAUS Jean-Marie
Risk Analysis: Socio-technical and Industrial Systems

FRIBOURG Laurent, SOULAT Romain
Control of Switching Systems by Invariance Analysis: Application to Power Electronics

GRUNN Emmanuel, PHAM Anh Tuan
Modeling of Complex Systems: Application to Aeronautical Dynamics

HABIB Maki K., DAVIM J. Paulo
Interdisciplinary Mechatronics: Engineering Science and Research Development

HAMMADI Slim, KSOURI Mekki
Multimodal Transport Systems

JARBOUI Bassem, SIARRY Patrick, TEGHEM Jacques
Metaheuristics for Production Scheduling

KIRILLOV Oleg N., PELINOVSKY Dmitry E.
Nonlinear Physical Systems

LE Vu Tuan Hieu, STOICA Cristina, ALAMO Teodoro, CAMACHO Eduardo F., DUMUR Didier
Zonotopes: From Guaranteed State-estimation to Control

MACHADO Carolina, DAVIM J. Paulo
Management and Engineering Innovation

MORANA Joëlle
Sustainable Supply Chain Management

SANDOU Guillaume
Metaheuristic Optimization for the Design of Automatic Control Laws

STOICAN Florin, OLARU Sorin
Set-theoretic Fault Detection in Multisensor Systems

2012

AÏT-KADI Daoud, CHOUINARD Marc, MARCOTTE Suzanne, RIOPEL Diane
Sustainable Reverse Logistics Network: Engineering and Management

BORNE Pierre, POPESCU Dumitru, FILIP Florin G., STEFANOIU Dan
Optimization in Engineering Sciences: Exact Methods

CHADLI Mohammed, BORNE Pierre
Multiple Models Approach in Automation: Takagi-Sugeno Fuzzy Systems

DAVIM J. Paulo
Lasers in Manufacturing

DECLERCK Philippe
Discrete Event Systems in Dioid Algebra and Conventional Algebra

DOUMIATI Moustapha, CHARARA Ali, VICTORINO Alessandro,
LECHNER Daniel
Vehicle Dynamics Estimation using Kalman Filtering: Experimental Validation

HAMMADI Slim, KSOURI Mekki
Advanced Mobility and Transport Engineering

MAILLARD Pierre
Competitive Quality Strategies

MATTA Nada, VANDENBOOMGAERDE Yves, ARLAT Jean
Supervision and Safety of Complex Systems

POLER Raul *et al.*
Intelligent Non-hierarchical Manufacturing Networks

YALAOUI Alice, CHEHADEHicham, YALAOUI Farouk, AMODEO Lionel
Optimization of Logistics

ZELM Martin *et al.*
I-EASA12

2011

CANTOT Pascal, LUZEAUX Dominique
Simulation and Modeling of Systems of Systems

DAVIM J. Paulo
Mechatronics

DAVIM J. Paulo
Wood Machining

KOLSKI Christophe
Human-computer Interactions in Transport

LUZEAUX Dominique, RUAULT Jean-René, WIPPLER Jean-Luc
Complex Systems and Systems of Systems Engineering

ZELM Martin, *et al.*
Enterprise Interoperability: IWEI2011 Proceedings

2010

BOTTA-GENOULAZ Valérie, CAMPAGNE Jean-Pierre, LLERENA Daniel, PELLEGRIN Claude
Supply Chain Performance / Collaboration, Alignement and Coordination

BOURLÈS Henri, GODFREY K.C. Kwan
Linear Systems

BOURRIÈRES Jean-Paul
Proceedings of CEISIE'09

DAVIM J. Paulo
Sustainable Manufacturing

GIORDANO Max, MATHIEU Luc, VILLENEUVE François
Product Life-Cycle Management / Geometric Variations

LUZEAUX Dominique, RUAULT Jean-René
Systems of Systems

VILLENEUVE François, MATHIEU Luc
Geometric Tolerancing of Products

2009

DIAZ Michel
Petri Nets / Fundamental Models, Verification and Applications

OZEL Tugrul, DAVIM J. Paulo
Intelligent Machining

2008

ARTIGUES Christian, DEMASSEY Sophie, NÉRON Emmanuel
Resources–Constrained Project Scheduling

BILLAUT Jean-Charles, MOUKRIM Aziz, SANLAVILLE Eric
Flexibility and Robustness in Scheduling

DOCHAIN Denis
Bioprocess Control

LOPEZ Pierre, ROUBELLAT François
Production Scheduling

THIERRY Caroline, THOMAS André, BEL Gérard
Supply Chain Simulation and Management

2007

DE LARMINAT Philippe
Analysis and Control of Linear Systems

LAMNABHI Françoise *et al.*
Taming Heterogeneity and Complexity of Embedded Control

LIMNIOS Nikolaos
Fault Trees

2006

NAJIM Kaddour
Control of Continuous Linear Systems

CPSIA information can be obtained at www.ICGtesting.com
Printed in the USA
BVOW08*2246291014

372843BV00004B/7/P

9 781848 214699